Praise For

THE
PERMACULTURE
Market Garden

Bringing permaculture's holistic thinking to the problems of market farming, Zach Loeks has done this burgeoning economic sector a world of good with his pithy words and lovely drawings. Not a book about carrots, but one in which vegetables, children, customers, trees, vision, and earthworms shape a matrix of success. Business planning was never before this colorful, soulful, or needed for the times.

— Peter Bane, author, *The Permaculture Handbook*

A commendable permaculture guide. If you're serious about generating a livelihood in partnership with the land, this book is your compass. Zach Loeks details an approach to informed ecological decision-making so close to my own work I can be sure this book will improve agricultural landscapes and lives.

— Jason Gerhardt, permaculture designer, Real Earth Design

A legacy of grateful land is surely a memorial that can capture both our imagination and our physical effort. Such a sacred mission deserves the level of thought Zach beckons us to pursue. Engaging in this discovery and planning process is worth the effort, and Zach gives us another tool to engage more strategically. Now get out paper, pencil, ruler, and graph paper. It's that simple, and that rewarding.

— Joel Salatin, Polyface Farm

A permaculture market garden is a living thing, with many moving parts, tied to the breath of the seasons and the pulse of the earth. To grow good food while regenerating the soil and nature is a noble endeavor. To do it profitably is essential for both the farm and the community it serves. *The Permaculture Market Garden* clearly demonstrates, with lavish illustrations and intricate details the processes of applying permaculture theory to the design, creation and management of these evolving systems.

—Darrell E. Frey, Three Sisters Farm, author, *Bioshelter Market Garden: A Permaculture Farm*

There are few books that delight the senses, satisfy the scientific itch and leave the inner environmentalist contented. I'm indebted to Zach for his work, not only for my own education, but for the many clients that seek a tome of collected wisdom as they embark into their own farming dreams.

— Javan K. Bernakevitch, B. Comn, All Points Land Design

Zach strikes me as the guy you want to spend a week on your property. Not just for his company, but to access his wealth of knowledge and experience in learning to understand the natural systems at work in any given landscape. *The Permaculture Market Garden* is playful in approach yet completely pragmatic in reasoning and methodology and should be seen as the permaculturist's guide to placemaking through Homesteading. This chewy tome of goodness will help you assess the ecology, geological history and potential future of the land you live on and work with. A wonderful guidebook to help anyone — with land big or small — to thrive not just survive, in tandem with their natural environment.

— Sharon Kallis, author, *Common Threads*

THE
PERMACULTURE
Market Garden

A VISUAL GUIDE to a Profitable Whole-systems
FARM BUSINESS

ZACH LOEKS

new society
PUBLISHERS

Cover design by Diane McIntosh.
Cover art by Jedediah Loeks.

All interior art © Jedediah Loeks.

Printed in Canada. First printing December 2016. Third printing May 2024.

Inquiries regarding requests to reprint all or part of *The Permaculture Market Garden* should be
addressed to New Society Publishers at the address below. To order directly from the publishers,
please call toll-free (North America) 1-800-567-6772, or order online at www.newsociety.com

Any other inquiries can be directed by mail to:
New Society Publishers
P.O. Box 189, Gabriola Island, BC V0R 1X0, Canada
(250) 247-9737

LIBRARY AND ARCHIVES CANADA CATALOGUING IN PUBLICATION

Loeks, Zach, 1985-, author
       The permaculture market garden : a visual guide to a profitable
whole-systems farm business / Zach Loeks.

Includes bibliographical references and index.
Issued in print and electronic formats.
ISBN 978-0-86571-826-5 (softcover).--ISBN 978-1-55092-620-0 (ebook)

       1. Permaculture. 2. Horticulture. 3. Sustainable agriculture.
I. Title.

S494.5.P47L63 2017   631.5'8   C2016-907513-3
                                                       C2016-907514-1

New Society Publishers' mission is to publish books that contribute in fundamental ways to building
an ecologically sustainable and just society, and to do so with the least possible impact on the environment, in a manner that models this vision.

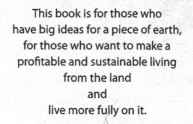

This book is for those who
have big ideas for a piece of earth,
for those who want to make a
profitable and sustainable living
from the land
and
live more fully on it.

Its concepts, designs and practices are:
Inspired by Nature,
Mapped like the land,
Goal-oriented for lifestyle,
Designed like ecosystems for ever-better production,
and
Managed for space/time/energy efficiency
for profit resilience.

Reader's Roll Call

Farmers, check
Gardeners, check
Ecologists, check
Educators, check
Future thinkers, check
Picked it up and it looked awesome, check

Foreword

Joel Salatin

Becoming acquainted with a piece of land is a never-ending journey, especially when you come to it with a mindset of receiving its subtle messages. As an early fan of permaculture, when *Mother Earth News* Magazine first interviewed Bill Mollison for the iconic Plowboy interview, I remember well the notion that nothing should be done to a piece of land until the caretaker walks it monthly for a year. That made a deep impression on me, even though I was just a teenager.

Of course, American sustainability laureate Wendell Berry and prairie iconoclast Wes Jackson both extol the virtue of knowing a place before working a place. All of these great evangelists of practical land stewardship presuppose a two-way conversation between the land and the caretaker.

Listening to the land sounds mystical and the stuff of fairytales, to be sure, but those of us whose life's work has been caressing a place know that this ear to the ground is the foundation of responsible stewardship. It's also the catalyst for massaging more out of the earth, as a lover and friend rather than conquered foe.

The more land under a person's control, the longer it takes and harder it is to discover all its nuances. Wet spots, dry spots, hot spots, cold spots, soil profiles and everything in between is a bit like the intrigue and sustaining joy of a good, healthy marriage. Although I grew up on our Polyface Farm and have lived here for nearly 60 years, just a couple of years ago I found a brand new spring nearly at the top of our highest elevation area, up in the forest.

On a late fall day I had gone up to this remote area to dream and think. The leaves were freshly off the trees, opening up the forestal understory to its winter nakedness. The setting sun at my back suddenly glistened on something through the trees that caught my attention. What was that? A piece of metal? Water?

Intrigued, I hiked down over the road embankment and stumbled upon a trickling spring. We were in a typical late-fall drought, but this spring, 900 feet higher than our farmstead house and outbuildings, offered beautiful clear water from a rocky mouth in the side of the mountain. Running at about 5 gallons per minute, this treasure was worth more than gold. Clean, high water is a godsend.

Some 700 feet lower we had excavated, over earlier years, a series of ponds to catch winter runoff. This water fed our 6-mile labyrinth of water pipe all over the farm. Gravity-fed, 80-psi water, no pumps, no electricity — true agrarian wealth. But suddenly here was a spring, 700 feet higher, better quality. My head buzzed with the possibilities: micro-hydro, irrigation, livestock water, potable water for agri-tourism cottages. Water is life.

Discovering the breadth and depth of your resource base is foundational to designing a synergistic, functional farming plan. Whether it's a backyard or million-acre ranch, land form precedes leveraging hills, hollers, soil types, gravity, and wind flow. What Zach Loeks offers in this little book with bold illustrations is a thought process to working with the land so it will want to work with you.

Most market garden books start with the plants, production techniques, marketing protocols and the like. This one dares to address the most basic climatic, topographical, even community nuances into the process. Fortunately, we have everything from soil maps to GPS to 3-D technology today to help us get better acquainted with the parcel of land we're getting ready to touch. Designing for efficient access, for micro-climates, for uphill-downhill flow BEFORE putting a spade in the ground can save us from years of frustration.

All of us who work with the land, be it livestock, produce, or crops, have a predisposition to impose something on the land that is not currently there. Rather than being inherently damaging, our presence can be generously helpful to the land. Our imposition can stimulate hydration, increase biomass production, develop guilds of symbiotic, complex relationships. Indeed, the human presence should and ought to be one that builds resiliency and more solar capture into the landscape.

In *The Permaculture Market Garden*, Zach gives us a pathway into this wonderful healing human presence. Rather than seeing our role as exploiters, he sees our role as friendly partners, teasing out of the landscape more than it dreamed possible in a static state. That's a powerful and ultimately hopeful idea.

So before putting spade to earth, before pounding corner stakes in to mark footers, and before calling in the well-drilling rig, take this book and work through it with your property. Use it as a template, as a planning guide, to build your dreams in regenerative systems. While it may seem anal and laborious to go through the process, the extra planning and humility, giving the land time to talk to you will yield a lifetime of benefits. Fortunately, the benefits are not just to you; they accrue to you because the land will be more resilient, more productive, and ultimately grateful.

A legacy of grateful land is surely a memorial that can capture both our imagination and our physical effort. Such a sacred mission deserves the level of thought Zach beckons us to pursue. Engaging in this discovery and planning process is worth the effort, and Zach gives us another tool to engage more strategically. Now get out paper, pencil, ruler, and graph paper. It's that simple and that rewarding. Go for it.

Joel Salatin
Polyface Farm

Thesis

A Proposal for Farm Profitability Resilience

Many of the points in this book are well known in the literature and to practitioners of the market garden trade. What I am presenting to you is a discussion that ties many individual components into a unified whole. The ideas are designed to advance the concept of *profit resilience*. I refer to profit resilience in the sense that farm profitability is derived from active support of *ecosystem services* through improving farm *regenerative productivity* now and into the future.

In other words, the means of production is regenerated by the mode of production. *Now*, because farm transitions need to maintain and even improve current profits and efficiencies. But more importantly, we must plan *into the future*, because productivity must marry sustainability, renewing the resources needed for production through the overall long-term process. For instance, market gardens should regenerate soil by increasing organic matter, biological diversity and soil aggregates through the very use of these soil assets in farm production.

Farms are sustained by the interface of earth, air, water and life; we must design production systems that replenish natural resources. First, we should *understand natural*

Resilience is versatile, beautiful, and profitable.

systems, so we can *design/manage* them by mimicking their regenerative nature. Active support of *ecosystem services* means directly sustaining ecological functions for improved agriculture. More than using soil to grow crops, it means supporting soil health for better production. More than farming the land, it means mapping resources with a focus on *holistic planning*. More than managing production just for profit, it means profiting because of better *agro-ecological practices*.

Yet, the very heart of successful farming is, as conservation efforts worldwide have shown, economically viable stewardship, namely practices that make sure those living on the land benefit more from better management of their prime resource — the land itself. Sustainability requires that productivity meet the needs of humans — albeit in a regenerative manner — for a healthy landscape is only maintained when those living in and around it are sustained as well. So, *economically viable regenerative productivity* must be the goal that we design for. This is profit resilience!

Profit resilience seeks an *annual return on perennial (lasting) investment* — benefiting *now* by investing for the *future*. Think about it for a moment: growing better salad greens because you planted a tree, improving harvest efficiency by maintaining cover-cropped alleys, improving garden yield with a hedgerow. We can invest in long-term production benefits (timber and nuts from planted trees), because we recognize other *hidden yields*, which benefit us in the short term (beauty, soil carbon, windbreaks).

However, these benefits from *perennial investments* (e.g., increasing soil organic matter) must be *valued and appreciated*. Putting value on unseen services of agro-ecosystems is paramount to a paradigm shift in agriculture. Widespread transitions to regenerative production (grass-fed rotational grazing for cattle, agro-forestry and permaculture market gardening) are amply justified when additional *indirect* returns (soil health) are also calculated in economic terms. When we consider the costs of fertilizer and the loss of it through run-off, then the value of cover crop is better understood. With this comes appreciation, a deeper respect for soil, which provides us with our breakfast, lunch and dinner. Thanks!

Organizational land patterning, using techniques such as *permabeds* (permanent raised agro-ecological beds) and *garden patterning* (framework for integration), contributes to efficient management of diverse annuals, perennials, animals and soil microbiology into a whole system. This uses *whole farm mapping* of farm property resources, micro-climates and natural flows to embed production in the local environment and transition land to chosen agro-ecosystems.

Holistic planning is a more comprehensive form of farm planning that focuses on the *farm family, community and the land* to sustain goals. Farms need a holistic goal, a guiding compass directing the care of people, community and land, creating a sounding board against which every decision is tested. We can also benefit from *ikigai* (Japanese concept), describing the individual's journey to balance: what you are good at, what the world needs, what you can be paid for and what you love. These ideas remind us to design the farm to allow oneself to grow in the journey through life.

Some additional components we need to incorporate for a *profit resilient farm* include:

Guild enterprise production, which helps integrate your farm's various lines of production, ensuring you don't take on too much and you work to build mutual benefits amongst your enterprises.

Brown brain investment is an annual investment in your soil and is consciously incorporated in your budgeting to ensure soil-building activities are sufficiently funded (and carried out) yearly.

Whole-system farm design places emphasis on the design process (observe, research, system design, action plan, act, monitor, observe) for all aspects of the farm so that they work together to enhance farm resilience.

Design management makes sure that our design process is living and working to effectively improve design over time. The understanding is that our design must be an evolving process, like an ecosystem, for success in agro-ecological production. For example, adopting design from other farms and adapting them to your context.

Permabed principles and production is the guiding concept for our permaculture market garden, including how we organize the land for improved soil health, ecosystem and crop services, and guild crop rotations that emphasize cooperation and biological mutualism between productions.

Agro-ecological landscapes are the overarching goal, the transition of our landscape to sustainable, regional food ecosystems of *regenerative productivity* for *profit resilient* communities.

Agro-ecological indexing is the process of introducing, trialing, monitoring and propagating species on farms and the sharing of the knowledge and genetic material outward. This culminates in the successful complement of these species soon defining the geographic boundaries of new agro-ecologies that are ecologically fit, production-efficient and commercially viable. In other words, successful diversified farms become pinpoints on the map testifying to these criteria and can be databased for a paradigm shift in production.

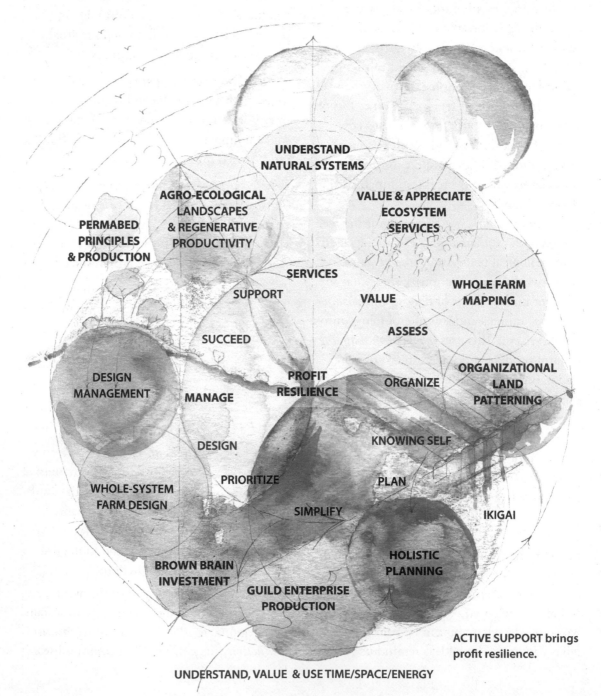

UNDERSTAND
NATURAL SYSTEMS

AGRO-ECOLOGICAL
LANDSCAPES
& REGENERATIVE
PRODUCTIVITY

VALUE & APPRECIATE
ECOSYSTEM
SERVICES

PERMABED
PRINCIPLES
& PRODUCTION

SERVICES

SUPPORT

VALUE

WHOLE FARM
MAPPING

SUCCEED

ASSESS

DESIGN
MANAGEMENT

MANAGE

PROFIT
RESILIENCE

ORGANIZE

ORGANIZATIONAL
LAND
PATTERNING

DESIGN

KNOWING SELF

WHOLE-SYSTEM
FARM DESIGN

PRIORITIZE

PLAN

SIMPLIFY

IKIGAI

HOLISTIC
PLANNING

BROWN BRAIN
INVESTMENT

GUILD ENTERPRISE
PRODUCTION

ACTIVE SUPPORT brings
profit resilience.

UNDERSTAND, VALUE & USE TIME/SPACE/ENERGY

Our Farm Is a Design Project

This book is for those who want to live on the land and make a living from it, who want to add to the soil every year under the shade of a tree, and who want to farm profitably, etc. I hope *The Permaculture Market Garden* inspires, informs and guides contribution to global agro-ecosystems.

Our farm is a design project. When we began tending our land, there was only a road. We observed nature over the course of several seasons while we worked off-farm jobs and operated a small market garden nearby. We observed the wind blowing, the snow drifting and where the ground was wet, dry and fertile. We let our land tell us its story and looked for opportunities within its ecology.

We mapped out a vision for our "back fifty" that included a market garden, heirloom seed production and an education/research center. We are there! It amazes me that we have materialized this destiny. Yet, as I write this book and bring together the core concepts, designs and techniques applied to our journey, I see it was no wonder we ended up where we intended. There were obstacles, but we stayed our course through compromise and design.

We worked consistently toward our objectives, defining what they were and how we wanted to get there. We wanted a sustainable and profitable farm. We wanted energy independence, a bountiful piece of Eden and meaningful work. We wanted to learn and teach, grow and celebrate the seasons. We wanted a good livelihood to provide for our family and fulfilling work to engage us.

So, we went off-grid, integrated animals, planted trees, grew crops and designed alternative systems, and we were…overwhelmed. We struggled with balancing diversity with the weight of responsibility. We had started farming with a bucketful of idealism. But then the load got lighter when we left some of it behind and simplified, systematized and prioritized. It became less about trying to manifest destiny, sometimes blindly, and more about understanding the farm life we really wanted and planning, designing and managing for it.

It is tricky to balance an ecosystem approach to farming with the efficiencies demanded of a market gardener. Yet, with a deep breath, we changed course and reevaluated what we were best at, what we wanted and how we might get there. Interestingly, as we simplified our approach to market gardening, innovating new techniques and gaining new perspectives, we ended up with a more ecological approach than we ever imagined. We also ended up going where we had originally intended — a successful permaculture market garden based on ecological production, research and education.

We trialed with error and improved with design as our new farm business unfolded.

A Hitchhiker's Guide to This Book

This book is a bit different than most of its genre. It is something between a graphic novel, owner's manual and scientific treatise. It is meant to be read from cover to cover, opened randomly and enjoyed, indexed in search of (or to revisit) a specific topic when needed.

This book has an introduction and then is divided into 6 sections. The first 4 sections build upon each other and culminate in the 5th section, The Permabed System. The 6th section has a glossary, bibliography and index.

The sections are designed to give an ever-expanding comprehension that starts with understanding natural systems and culminates with the indexing of agro-ecological landscapes. This journey is holistically briefed in the *Profit Resilience* image (page x) and as a thesis titled: A Proposal for Farm Profitability Resilience (page viii). It also follows a continuum of farm design for permaculture market gardening and sustainable agroecological farming (page 6).

This book makes ample use of both words and images, complimenting text with full, half and quarter page watercolour designs and illustrations. It is written in a more dialogue format with a stream of visual boxes, including *Green Thumb Techniques*, *Do You Know* boxes, *Principle Spheres*, *Design Tips*, and so on, to provide tangents, details and humor for the overall discussion.

There is a lot of terminology, some new to readers, which appear capitalized and/or in italics throughout the book depending on their significance and will, depending on the need, appear in either the glossary, be illuminated within the text or found in a Jargon Margin at the bottom of the page.

Because this book speaks to a more holistic approach to farming, terms and concepts are often referred to before they are expanded upon. As such, make use of the glossary and note the bracketed references to the section where concepts are expanded upon.

This book is designed to encourage more creative thinking about how we design/manage our planet Earth. It is a call to arms for success to each in their own way. As such it presents a great array of material and I hope readers will pursue some topics in more detail at their leisure.

Get 'er done!

This book uses colors to conceptualize that there is more when elements come together because in nature, as on the page, profound secondary colors are created from the primaries.

Contents

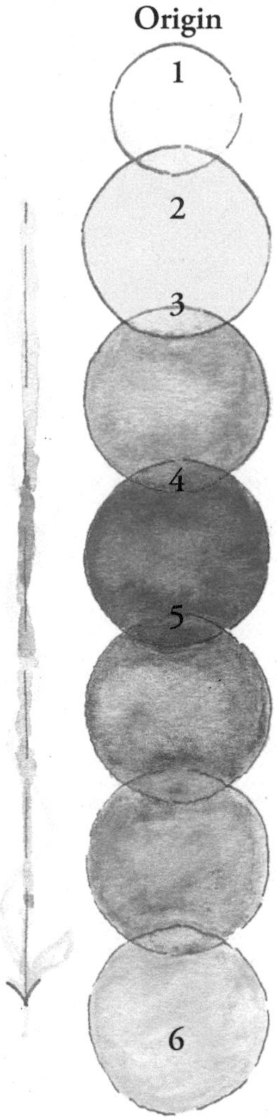

Origin
1
2
3
4
5
6

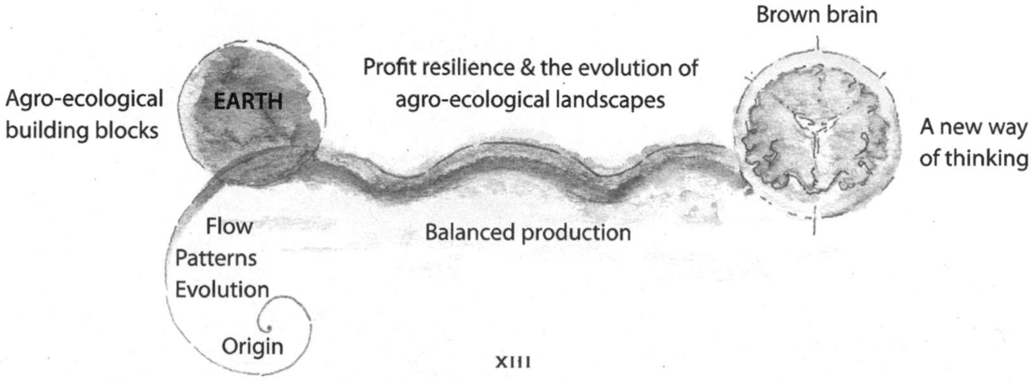

Brown brain
Agro-ecological building blocks
EARTH
Profit resilience & the evolution of agro-ecological landscapes
A new way of thinking
Flow
Patterns
Evolution
Origin
Balanced production

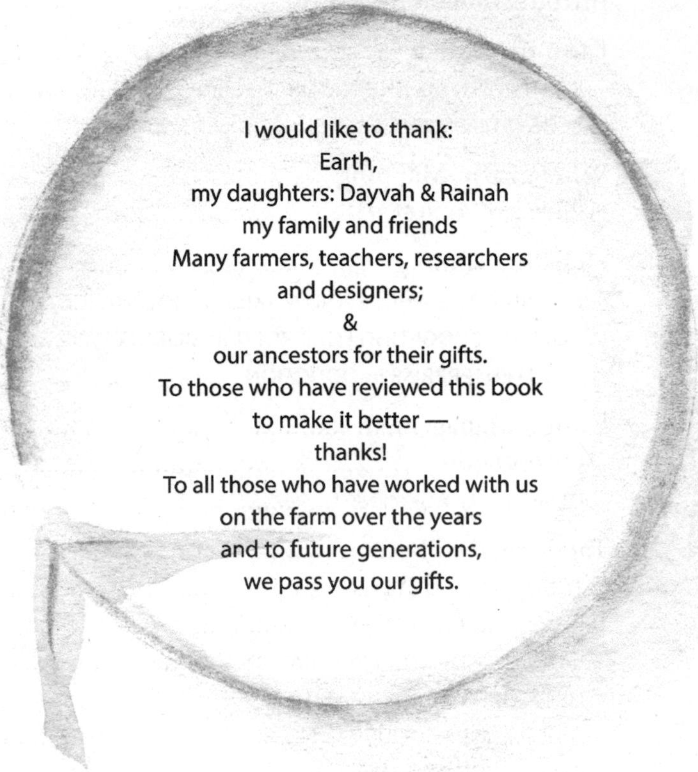

I would like to thank:
Earth,
my daughters: Dayvah & Rainah
my family and friends
Many farmers, teachers, researchers
and designers;
&
our ancestors for their gifts.
To those who have reviewed this book
to make it better —
thanks!
To all those who have worked with us
on the farm over the years
and to future generations,
we pass you our gifts.

A Place in Space

4.543 billion years to make this precious resource.

Introduction

Agriculture: Now and Future

Agriculture can be very destructive, often using a disproportionate amount of energy (resources, labor, fuel and money) to produce food energy (calories, fats, vitamins, etc). Furthermore, large specialized operations are inflexible to change. Overinvestment in heavy equipment to produce a few cash crops leaves the farmer vulnerable to global market trends and natural phenomena.

Conventional systems are often shortsighted and miss the horizon: ecological farming is more profitable. It is resourceful (able to use less time, energy and space), resilient (meet economic and environmental change through foresight and design) and regenerative (farmland begins to be self-maintained for healthy food ecology). It also produces a better product (tastier, more nutritious), sustains local economies (small farms are apt to hire and purchase locally) and asks a better price for the farmer for its inherent value.

On one hand, annual cropping is continuously dependent on inputs and maintains vulnerable field environments that are ill-equipped to help crop production. Crops are small and fragile in a wasteland of turned ground, which, exposed to the harsh winter, is less suitable for soil life survival. In this model, expensive fertilizers are required, yet they run off the overtilled and compacted soils.

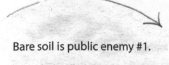

Bare soil is public enemy #1.

Living soils can better cycle nutrients, and integrated polycultures produce material for soil building and sequester and fix nutrients, for cycling to crops.

Ecosystem-based production can focus on diversity and soil health, integrating annuals with perennials and including animals or other species in the rotations. These agro-ecosystems focus on maximizing the exchanges between diverse facets for the overall benefit of the production. They are time, space and energy efficient. For instance, a crop's needs (fertility, water, heat, shelter) might be produced in situ, found as a side benefit of a complementary production, maintained through a better investment strategy for crop care or a more holistic approach to crop health.

Why don't we grow our own tomato stakes? Right in the garden! Coppiced rows of trees within our gardens could provide windbreaks, snow fences, summer shade for spinach and trellis support for grapes and also be thinned for tomato stakes. Instead of purchasing these stakes, we annually harvest them from a perennial bed integrated within our agro-ecology.

Again, more common agricultural models are yield-obsessed. This constant push for highest annual yield dictates farm management, losing sight of the many *hidden yields* and services provided by the farm ecosystem and its crops. These are often unnoticed and unvalued. These are often multifunctional, hence, relatively free! It is design and management for these side services that is needed.

Innovation for the future can include ecosystem services integrated into farm infrastructure. Take, for instance, capturing and storing water for irrigation to hedge against unpredictable droughts. Or consider our root cellar, where we use the free geothermal heat to keep our vegetables stored well through winter and fill it with cost-effective ice for summer cooling.

Do You Know? SPIN Farming (Small Plot Intensive Farming) is a burgeoning movement in urban areas making use of precious lawn, balcony and park space to produce local food.

Greening Cities

Agro-ecological cityscaping
SPIN Farming
Urban Gardening

The New Green Belt

SPIN Farming
Green Parking (P)lots
Edible Landscaping

Townside

Town-side Market Garden
On-farm pick your own
Agro-tourism

Countryside

Field-scale Market Gardening
Laneway Farming
Guild Cash Cropping

Wild Country

Sustainable Forestry
Integrated Edible Forest Production
Rootcellaring

Wilderness

Inspiration
Research
Wild Genetic Diversity

Food is produced at every scale, and we require patterns for each.

New Farm Rising

North Americans are left short-handed by the loss of a well-rounded diet. We eat primarily corn, soy, sugar, wheat and meat from animals fed a similar diet. Affluence is becoming a struggle against obesity, depression and boredom. This results in malnutrition, poor work conditions and loss of traditional farming in many communities.

We need to focus on growing the food that everybody needs where they need it. Sustainable food production is needed at every scale, feeding the communities around them and generating meaningful trade with

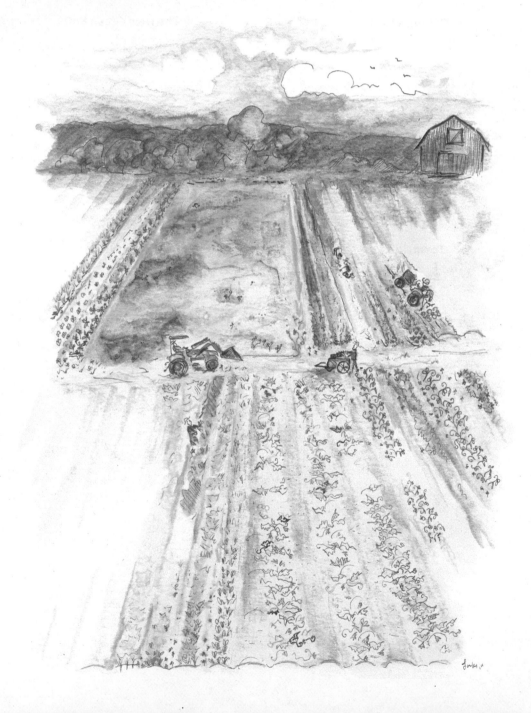

other countries for that which we cannot produce. In a modern society struggling with unemployment, creating livelihoods in such meaningful work as food production is priceless.

However, this seems easier said than done. The way we produce food is neither truly profitable to people nor sustainable in the long run. Corporations sequester wealth, the planet is stripped of diversity, and many farmers make little income, depend on crop insurance and subsidies, and retire by selling their farms. This doesn't seem very attractive for a new generation. Indeed, most farm kids leave the farm after seeing their parents struggle.

Yet, the small-scale farm movement has provided a new framework for profitable farming. It is sustained by production of high-profit crops sold direct to consumers, specialty stores and restaurants. It is centered on increasing diversity. This is great! Its diversity and hands-on approach multiplies the potential ecosystem services and provides many niche crops for your community.

So attractive is this bright new future that the youth of today are flocking to backyards, urban lots, suburban lawns, rural fields and mountain nooks to build their dream of successful farm life. Never before has the ground been so fertile for a generation of entrepreneurs willing to get dirty and live on the land. There are countless educational and farm-start opportunities: WWOOFing, internships and incubatorships. Many organizations offer grants, idea sharing, and advocate for agricultural change on many scales. There are new farms rising!

However, the local food movement is still largely dependent on short-term profitability, limited diversity and excessive energy inputs. Market gardens need be wary of falling into the same pitfalls as broader agricultural production. We seem to limit our crops to a specific range of annuals or a handful of

Do You Know?

WWOOFing or Worldwide Opportunities on Organic Farms is an organization linking host farms and volunteers in a cultural and educational experience to build local and global communities around food and people.

perennials, feeding them excessive amounts of on-demand energy inputs, overtilling our soils and overworking our souls. Most are dependent on the sale of a few crops and go to all extremes to produce these in quantity and quality in a monoculture field. I get it! I have done it.

But the issue is more complex. For, on the other hand, farms can become too alternative-minded and idealistic, diversifying to no end, unfocused and spread thin. A farm cannot attain profit resilience or protect the environment if it is too unfocused to successfully nurture and manage its enterprises. Many young farmers burn out before they can become truly profitable. True profit covers cost of production *including* a living wage for the farmers.

Yes, a new era in farming must come to fruition. An era where we build more soil, store more water, grow richer foods and support happier communities. Yet, a practical approach is needed to manage these ambitions, and fluid transition is paramount. One that moves toward ecosystem farming without interrupting the needed cash flow of current farm operations. One that emphasizes simplicity of management alongside complexity of ecology. One that takes the farmer's health and livelihood in hand.

The Permaculture Market Garden is a road map for new and established farms toward profit resilience, true profit, one that doesn't jeopardize future profit by gnawing away at its foundation, like the eroding soil beneath your feet. Profit resilience is the annual return from perennial investment. Investment in soil, in ecology, in better crops, in family livelihood; investment in the now and the future together; in systems that mimic an ecosystem so we can be profitable and resilient like one.

> *Profitability requires that we make more money than we spend, profit resilience requires that this be sustainable. Sustainability requires that we work within natural constraints.*

The Journey Forward

This is a journey. A journey of self-reflection, natural observation, priority planning and better design. A journey of management, of systems and solutions.

A journey that starts by understanding natural systems, by valuing and appreciating what they have to offer.

It is a journey that uses mapping techniques to better see what resources and realities you have on your piece of land. It is a system of organizing the land around its

Our cycle of farm design

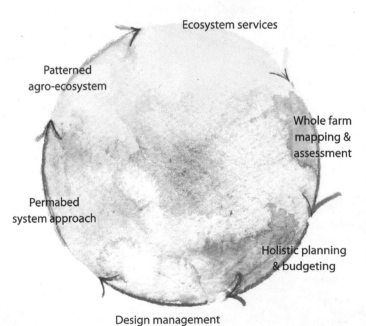

Ecosystem services

Patterned
agro-ecosystem

Whole farm
mapping &
assessment

Permabed
system approach

Holistic planning
& budgeting

Design management

nature and for your management. Where the wind blows and the water lies.

It is the process of planning your farm with you, your community and your land in mind, of knowing where you want to get to and making sure you put in place what is needed to arrive.

It is a way of simplifying your production so it is efficient and still has all the benefits of diversity. Focusing your management so you can succeed at what seems a daunting task, managing an ecosystem.

This is a journey of designing and managing agro-ecology for profit. For without profit, your farm cannot succeed.

You must pattern your production for ecosystem services to truly realize the highest potential yield from your piece of land now and into the future. Integrating cover crops into your garden (same plot, same season) is an example of patterning for ecosystem services. The cover-cropped bed can act to shelter some crops, become mulch for others and further decompose to be food for yet another. It also serves as habitat for an ecologically complex gardenscape.

This journey is about putting in place a system that will benefit many generations to come, and yet — here is the clincher — they must also benefit you now. You must see return on these systems as you would any other investment, for then they hold their own and can stand up and shout out, follow me to greener pastures and lush hedges, to fuller fields and happier heifers, to healthy communities and cleaner ecologies, for I will help you build a farm that is beautiful, sustainable and also profitable.

Turn-of-the century farm from my grandfather's sketchbook.

SKETCH BY CONRAD DAVID LOEKS

Where Is Agriculture?

Beginnings of Sustenance

Food sustains us. Where and however the food grows, it does so because of sunshine, earth minerals, water flows and life that grows.
A little perspective, humility and knowledge can help us along the way to better agriculture. We need not lose track of profit, but only remember that that which sustains us now and into the future is actually Earth. We are actually part of Earth.

Natural elements

When we step back and look at the evolution of our planet, we are humbled by the reality. Our Earth formed around 4.5 billion years ago with modern man only emerging 35,000 years ago. Agriculture, with domesticated species, has been around for a mere 10,000 to 15,000 years, and the first metal plow was used in the 1700s. Modern globalized agriculture and agri-food businesses that dominate current food production arose in the 20th century. Only over the last decades of the 20th and early 21st century has modern organic farming been gaining momentum. We should look to the future and ask ourselves: what horizon are we shaping? How can we learn from the past?

Our earliest ancestors depended on the cycles and systems of nature. Intimate knowledge of nature rewarded us with food: fresh berries, roots, nuts and meats for hunter-gatherers. Understanding of weather, ecology and geology allowed early humans to develop revolutionary techniques and technologies: seasonal migration to areas of abundance, use of fire for heat, cooking and land clearing and eventually the accidental — followed by intentional — selection and domestication of plants and animals.

Modern society is a product of the transition from hunting and gathering to sedentary farming. Our ancestors worked the land, year after year, turning seed, water, soil and sun into food. At first we moved from place to place, slashing and burning new plots of land and enjoying the nutrient-rich ash- and weed-free conditions, before moving on. Soon, we settled areas, preferably with rich soils, clean water and fair weather. We built a whole new way of living, with houses, roads, fences, government, trade and money. A new specialized way of living based on an agriculture that created surplus.

Where Is the Agri-culture?

Unfortunately, this specialization has led to the disconnection of people from food. We now live in a society where the average person doesn't know that broccoli is the edible budding head of a plant. Why should they? They don't grow it. They make money and buy it. We have specialized to the point that many don't even know those who grow it, or where it grows

A culture of food vagueness. This is culture without agri…

Yet the more devastating conundrum is that even the farmer is losing connection with the land. Our rural landscape has transitioned away from the mixed-production family farms that used to be so prevalent, to large monocultures, factory farms and agribusiness. I once met a lady who told me she grew up on a farm in Ohio. She still owns the land, but she has no involvement now. It is part of a greater network of acreage farmed by a corporation. No one lives on the land, but many employees come and go and work it with machines. She didn't know what crops were even grown. "Probably soy," she said. An agribusiness without intimacy. Disconnetion.

This is agri without culture.

Without families owning, living and farming the land, we will lose the stewardship and care necessary to create the sustainable farms of the future. When you live on the land, you are more likely to care about the long-term health of the soil, the cleanliness of the water and the community around you. If you live where you farm, you have to live with the repercussions of your practices. Away from the land, it will be difficult for farmers to know it intimately and build strategies for sustainable production that work with the cycles of nature.

Diverse market gardeners have a remarkable opportunity to invigorate *agri*culture.

They can make good profit through direct marketing and value-added products. This allows them to work intimately with land and its cycles. They can stay small and resilient in the face of change and build systems of improved production, grounded in an ethos of natural observation, understanding and appreciation. This is balanced by focused production, goal-oriented decision-making, and sound management techniques. Farming more like we once did, but with an eye for the future.

Building Intimacy with Nature Through Design

We need to relearn the intimacy we once had with nature. As farmers, we must learn about the natural systems that surround us so we can better care for our crops. We too often forget that fighting nature is an uphill battle, and it is worth understanding those most powerful entities, wind, water, sun, soil and biology, and working with them. Watching their patterns to mimic their services. For instance, a weedy bed of germinating lamb's quarter can be grow as a green manure; it is quite good at accumulating macro- and micronutrients, without the normal monetary or labour cost associated with cover cropping. Understand the plant's life cycles and mow it as a green manure before it seeds. As we will discuss, a system of flexible garden management is the crux of how we can better make use of natural services without a logistical nightmare.

Let's design natural phenomena to be integrated into the farm. Wind can blow a greenhouse away or prevent mildew, it can dry and cure your onions or whip your transplants about and dry their roots. Consider nature and build intimacy.

Habits Destructive to Sustainable Farming

Many market garden techniques are currently destructive to the land. A clear example is ubiquitous reliance on tillage for land preparation, weed eradication and pest management. Millions of tons of topsoil are lost to wind and water every year. This is an essential and irreplaceable ingredient for food. Our current practices need improving.

Not only do these techniques destroy the soil we depend on, but they are far from the most profitable approaches to growing food, ultimately costing billions of dollars in fertilizers, pesticide and other inputs. Alternative

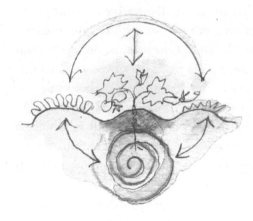

tillage, soil life benefits, cover crop methods and polyculture production are a few approaches inspired by nature that can help the transition to profitable market gardening.

Weather Management

	Sunny/Hot	Sunny/Cool	Cloudy	Rainy	Windy	Frost	Adverse Weather
AM	Tractor weeding	Hand weeding	Harvest	Observation	This is not good weather for laying row cover, transplanting bare root plants or building high tunnels	Protection	
			Transplanting				
		Row cover		Barn work			
		Seeding					
PM	Land preparation			Greenhouse work			

A good example of design management is managing for the weather. It makes no sense to work your hat off in the wind putting up a high tunnel and then to go inside on a calm day and organize the barn. Don't fight the weather; manage for it.

These approaches are far from new, having roots in our ancient past, and emerging again and again throughout the history of farming. Most recently we have seen them emerge in the last century in response to the dust bowl and the green revolution. Work by individuals like J. Russell Smith with his seminal work *Tree Crops: A Permanent Agriculture* and organizations like the Rodale Institute with the research into no-till agriculture and polyculture indexing by David Jacke paved the way for modern sustainable farming.

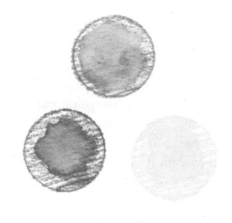

Simply Integrate Productions

Indeed, if I say anything, it is that we should integrate perennial food productions with sustainable annual crop production and rotational animal systems. This is inclusion of nature in food production in an organized and profitable manner.

That complete and utter diversification should occur immediately on every farm is a far cry from the reality of what it is like to manage a farm, because a farm needs simplicity, efficiency and order to be manageable. But, there should be *more* diversity integrated on every farm and diversity structured to improve each other's production through a supportive system such as guild-based agriculture. Efficient guild crop rotations between permanent perennial beds create focused complementary diversity that is ecologically appropriate to the land and considers the farmer's current skills, tools and goals alongside community demand.

When you step back and look at monoculture production, deliberate and incessant soil tillage and heavy reliance on pesticides, it becomes clear that the agriculture of today is out of perspective. Nothing like this occurs in natural systems. Not only are we fighting natural order, but also we are failing to receive its services. Layering production through use of trees, shrubs and annuals; receiving fertility from living animals and using conservation tillage techniques that leave the soil profile intact are all positive steps forward.

It is paramount that we develop techniques and tools that allow market gardeners, in their current situations and landscapes, to transition slowly and effectively without reducing income while increasing profit. Movement toward permaculture farming should be profitable.

My Story

As a child, I planted strawberries under my kitchen window, where they grew plump against the sun-warmed stucco wall. The cherries grew up a trellis against our utility shed, and large apricot, plum and apple trees spread their branches to the blue sky. Sheltered from the winds by garden walls and fences, a small oasis of mulched annuals grew, fed from rain barrels and desert rains. I took for granted that I grew up on a permaculture homestead. I just thought the fruit was tasty and climbing the wall was a great way to get at it!

I remember transplanting trees with my dad and brother on properties in northern New Mexico. We planted for future yields, dug swales to hold water for the plants and built *straw bale berms* to counter soil erosion. We also spent a lot of time running around nurseries, learning the names of strange plants and bringing them home to grow.

As a youth, I traveled to India and saw the life spilling over on every street. The sheer numbers of people opened my eyes to the hunger of our world. I began to see some of our planetary plight and how little so many had. Yet, in a small mountain village, an eight-year-old girl took me along a mountain path and pointed out every plant we came upon by name. She told me all the healing each plant provided. She didn't know the precise properties, the science, but understood all the ailments each could alleviate. She showed me how much could be had from our Earth.

Excerpt from India journal

As a young man, I studied geography at Concordia University in Montreal. I hung on the words of professors and delved into the books, learning about our planet, its ecology, climatology and geology. I learned about humans and our growing civilizations. I embraced conservation and resource management and looked critically at our natural world and the resource-hungry humans in it.

I remember a professor saying we will never solve our planet's problems by excluding people from its resources. The success of land conservation wasn't through exclusion but inclusion. When people in an area made a good livelihood through ecologically responsible jobs, then both people and planet thrived. How can we create more ecologically responsible jobs?

Chimayo, New Mexico. Home of its famous namesake, the Chimayo chili pepper.

ILLUSTRATION BY CONRAD DAVID LOEKS

Now, I realize that the same goes for soil and food production — we need to re-immerse people and families into our food production landscape, not just abandon it to corporations. Without people there is no concern for the future yields, and without profit there will be no people left to care.

Over my academic years, many frightening facts furrowed my brow. I was told that today when we supposedly know better, areas twice the size of Rhode Island are being cut down in the Amazon every year to plant soybeans. Humans are cutting down some of the Earth's most productive polycultures to plant a monoculture. This struck a chord in me. Didn't we need this diversity? Weren't the cures to cancer and the next disease-resistant food crops out there? Was that not the primordial feast of our ancestors?

It was at this time that I read *Historical Geography of the Americas* by William

Cronon. He described the forests that existed in North America when Europeans first arrived. Oh my! Where are the great food forests that used to spread across North America and other continents? They sounded wonderful, and I was drawn to the deep forests for answers. I traveled north into the boreal forest to live amongst the trees and learn from them firsthand.

My wife and I traveled to northern Quebec to conduct forestry research, studying the regeneration of the forest after human and natural disturbance. We looked at alternative harvest techniques and the interactions of fire with the ecosystem. It became clear to

me that forestry may be losing sight of the profitability of natural forest regeneration. Fire was providing an important ecosystem function where the parent trees opened their cones under the presence of fire, releasing their seeds down to the freshly burned ground where new clean seedbeds awaited. The seeds flourished in these nutrient-rich and competition-free seedbeds. Quickly a youthful forest grew where the burned forest had stood. However, natural regeneration can take time, not all species are desired for wood products, and the future forest would be less orderly than the planted. Order is always desired by the human mind. Is there a middle ground? Are there other services and products we should consider?

The current practice was to clear-cut all the burned forestland, leaving the opening cones in big slash piles where the seed rotted. Instead tree planters had to physically replant the forest with plugs brought in from nurseries in southern Ontario. These resulting forests have narrow genetic diversity and less local adaptation. These forests may be more prone to epidemics, such as insects and disease, and less hardy to weather extremes if grown from southern seed banks. Even when genetics are known, we need ask: what is lost? Yet, climate change may shift the boundaries

of bioregions quicker than trees have historically adapted and importing southern genetic stock may be important. We should prepare to tend our global forest for the future climate.

Organization of forested land could lead to improved diversified production for northern communities. But it must be done thoughtfully.

Fire in the boreal ecosystem was evidently important for establishing stands of healthy and diverse tree species. A young forest of germinating spruce and pine, bright green against the ash black, is a well-designed sight to see. But salvage immediately after fires has become increasingly popular as a way of making the most of the valuable timber lost to fires. Unfortunately, when hastily done, it results in erosion, loss of important habitat and reduces this key service of natural regeneration. So, our research considered the outcome of leaving a percentage of the burned trees standing as parent trees for natural regeneration. Can we afford to value the service of a tree left standing?

As I returned south to Montreal with the Beatles' rhythms on the radio soothing my soul, I realized that nothing is as simple as seeing nature working and saying we should just let it be. We need to find ways to work within the system but out of the box. Find profitable alternatives to current practices. Find alternatives that are easy to transition into; nobody wants to be uprooted and have everything drastically shifted. But everybody will make little changes if they see that it is profitable and also sustainable.

We returned to the family farm and threw ourselves into the reality check of owning and operating a business. Driving past fields of corn, soy and wheat, we started an ambitious venture to grow what people eat. We started a farm and began to grow vegetables for a local market and small CSA (Community Supported Agriculture). Learning to grow vegetables we also began to appreciate the difficulties of diversity. It was more work than our idealized dreams.

We were growing corn, lettuce, basil, cabbage, celery, leeks, beans, radish, potatoes, onions, squash, etc. It was hard work! Each crop had its own needs for fertility, pest management, harvest and storage. When we expanded our gardens, now growing many varieties, I remember observing other vegetable farms growing only five crops. I started thinking maybe that was the way to go. Less is more, right? Wouldn't we be better off if we had fewer types of pests to worry about and simplified the growing conditions, harvest techniques, storage needs and marketing?

It was at this point in my farming career that the memories I carried from my childhood returned to me. Those memories of growing up on a small permaculture homestead in Santa Fe. We always had food aplenty growing around us. We did plant, weed and water. Yet somehow our small urban homestead in the mountain desert seemed to efficiently yield every year. We had apricots, plums and cherries, vegetables, sheep, chickens, goats and rabbits.

We were growers. But growers of guilds. We grew corn, beans, potatoes and squash. But we grew them together! We didn't separate everything out but grew it together, and I began to think about reintegrating my

crops into groups in the fields more like we used to grow. Also, we didn't always do the same thing every year. Some years we had a flock of sheep and some, a big pumpkin patch. But we always had our staples, our perennials. Is their value in this? Was this the direction to go?

Immediate Return on the Future

My two favorite places, the garden and forest, were becoming monocultures around the world. This was crazy since humans enjoy

In-town opportunities

a diversity of plant and animal foods and products. Yet we seemingly refuse to integrate them. So, despite my urge to simplify, I began to look for efficient ways to further diversify with trees. I did this despite the knowledge that the cost was up-front and the return was further down the road. Or is it? Is there not immediate value from planting a tree? Beauty, windbreak, shade.

Perennials proffer up riches annually and more immediately than we realize. Consider the services of an immature tree, the way it shades the soil adjacent, holds the ground and sheds leaves; harboring soil life amidst its roots and feeding organic matter from its leaves. The soil this creates, the complex aggregates that bind nutrients and the cycling to neighboring plants via symbiotic mycorrhizal fungi protected by that very same shade.

What is the ROI (Return On Investment) of a forest market garden? Or ecosystem market garden? Of an agro-ecology? It could be forest or grassland, or better yet a woodlands (with very high net primary productivity), if we are truly trying to maximize photosynthetic gain. Sounds reasonable to me. After all, as farmers, we are in the business of turning sunshine into products; maybe we should enhance photosynthesis.

What is the ROI of planting trees in your garden, along your fences and drives? Of integrating and diversifying your annuals in an organized manner? Of fostering relationships with soil life and seeing the hidden yields and services of the *farm ecosystem*? Fruits and nuts and berries, herbs and vegetables, pigs and chickens, mushrooms and meditation. Resiliency in the face of change; forget crop insurance, I have it other ways.

Meditation, yes, for not only did these aforementioned gardens of Eden provide bountiful food, they also gave contentment to our yearning souls. I don't know about you, but the initial joy of working in a field quickly wears off when you are toiling all day under the sun tilling up earthworms and watching their halves wiggle in the sun. Where is the shade of trees with fresh fruits? Pick one, take a break under the shade and feel peace of mind knowing the soil is alive, well and doing work for you.

A New Agriculture for Our Farm

It was in the midst of a 50-year drought when we were working twice as hard for half as much that I knew we needed new agricultural solutions for our farm, and the churning cauldron of design began to simmer, producing ideas. Ideas for transitioning toward profit resilience. A model where every year we conserved more water, planted more trees and enriched our soil with the life that drives its mighty engine. A model where the efficiency needed on a farm wouldn't be compromised, while maximizing the resilience and profit potential of the farm through diversity. We would work to create a farm for the future. A farm that would be designed to let nature work for us and let us work for nature. Holding our future in common. And so a new pattern began to unfold across our land.

Now my family farms this land. We farm it for the future. Our living connection to the land promotes best management practices, less pollution and more long-term sustainable investment.

So here is where this book starts. This book is about planning and designing small and medium-scale permaculture market gardens. It is a dialogue on ideas for the future of market gardening. It is geared towards moving toward efficient commercial permaculture systems.

I never thought I would be a farmer. I always thought I would be a writer or inventor … or a volcanologist. I guess I have come full circle. In permaculture farming, I have found a way to express my creative energy, innovating and keeping my hands in the earth: building, growing, designing from this piece of land on planet Earth.

I am glad to share through this book.

It is not the perfection of natural farming that we should strive for, rather the journey: practicing each day the efficiencies, profits and well-being of working alongside natural systems to produce food.

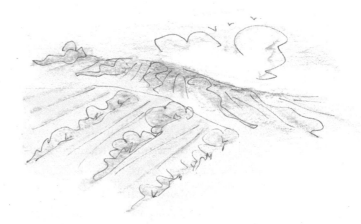

For the most benefit from this book, read it from cover to cover, think, research and read it again. Then design ideas into your farm and redesign so it works for you.

The following list will help you grasp the major concepts of the book.

This Book in a Nutshell

If I said nothing else, I would say this.

- Understand natural systems: the hydrological cycle, soil food web and plant life cycles. Understand how these interact on your farm.

- Make use of ecosystem services. Trees stabilize the soil, provide organic matter and shelter crops from wind and sun. Soil organisms help cycle nutrients to crops and build healthy soil.

- Understand that even *your crops* have their own services above and beyond the food you harvest; learn these by appreciating their full life cycle and design to maximize them. Peas fix nitrogen and squash shades out weeds.

- Implement long-term investment in soil through amendment and conservation. This means adding good stuff like mycorrhizal fungi and protecting soil structure and organisms.

- Value and appreciate ecosystem services and consider their monetary worth to your business.

- Map your farm, become familiar with its resources, ecologies and microclimates and make use of these where practical.

- Organize your land with definable units, be these beds, fields or paddocks that help you integrate productions and define management areas.

- Turn space into pattern units for design and management.

- Integrate projects, processes and products in a given space, be it a field plot or a wash station. *Manage for improved flow, exchange and feedback.*

- Make a holistic goal for your land, family and business. Clarify your long-term farm goals. Make sure to farm the farm and not let it farm you.

- By knowing where you want to go, you can plan to get there profitably. There is nothing profitable or sustainable about a farm enterprise that keeps you from the lifestyle you want or drains your resources for some ideology.

- Design for multiple functions for all supplies, tools, buildings and people. Make the most of what you have and use it more fully.

- Design for extended flexibility so your business is adaptable to change and versatile for various futures. The bigger the investment, the more versatile it should be.

- A decision delayed is a decision better made. Consider it, walk your land again, sit in a high place or lie low amongst the rows of peas and gain perspective.

- Don't put all your eggs in one basket and have more than just eggs!

- Build your farm around core profit centers that are self-sufficient but mutually beneficial. Three centers of production that can build into three balanced farm enterprises.

- Use the Rule of Three to limit enterprises as it minimizes management complexity yet allows a great diversity of production.

- Make these enterprises mutually beneficial like a plant guild.

- Don't produce so much diversity that you have no clear products and little efficiency (the produce *everything* idea is crazy). The many benefits from diversity are lost if poorly managed.

- Invest in soil improvement annually.
- Put all your garden into permanent raised beds.
- Integrate perennial and annuals within a garden in a distinct pattern.
- Get a notebook and use it on a regular basis to take good notes of production, experiments, seasonal phenomena, etc.
- Split the book in half: one side for notes, scribbles, calculations and the other space to come back and think over your notes critically and make actionable comments and conclusions.

- Adopt systems and designs from other farms and then mold them to your context and adapt them to suit your operations and lifestyle. Don't blindly apply another's approach. Consider why it works on their farm and critically evaluate how it might work on yours.
- See how your farm fits into the local, regional and global movement. What is your research contribution?

Farm Design Cycle

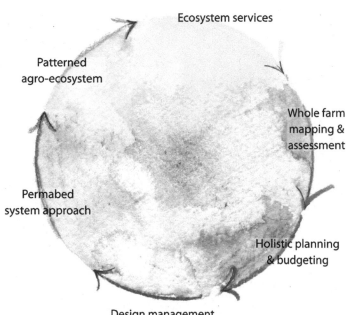

Ecosystem services

Patterned agro-ecosystem

Whole farm mapping & assessment

Permabed system approach

Holistic planning & budgeting

Design management

Farm Ecosystem

The Study of Our Systems

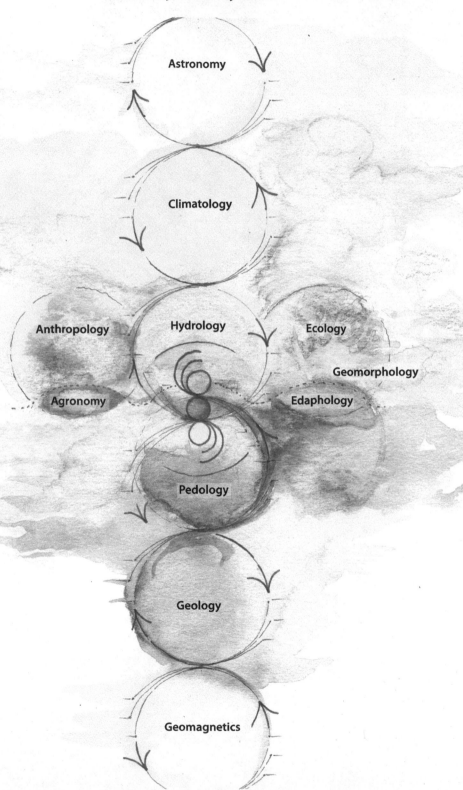

Earth Systems and Natural Science

A Natural Truth

We Must Start with Nature

Socioeconomics and culture exist within natural systems. Humans have studied the natural world and organized it into spheres, systems and scientific laws. Our mathematics, philosophies and politics are derived from nature. There is a greater truth out there. A natural truth. *The truth that water flows downhill.*

When I was young, my father, a permaculture designer himself, showed us this truth through trips into the backcountry of northern New Mexico.

Many times when camping in the mountains, we would play amidst myriad creeks that fell sharply from balsam- and spruce-covered granite ridges to the valleys where the aspen grew. Here the water meandered tightly on course to bigger waters; when obstructed by roads, it followed ditches.

Watching the waters flow, we made bridges and fish weirs out of the red and black stones that had tumbled down the weathering hillsides. We cheered and played "One, two, three! More! Here, fill in some more here, there are some little fish there!" My brother and I worked diligently. We fortified against the flow, the water slowed and deepened and eventually seeped over the sides.

"The water is moving, watch the way it turns!" my dad commented. We watched the water swirl to make way. Fish forgotten, we delved into the pure mastery of nature.

"More rocks?" my sister was determined to aid us in the exploit.

The water was constant as we interacted with it. It moved around our stones, rose and flowed over, squeezed between and eroded material out of the way. As the slope of the land grew steeper and small streams coalesced, the water was working under greater forces, unwavering, and becoming part of a larger system.

We grew exhausted in our labors, our original goal, the provisioning of food for our camp, forgotten. Let us not forget the aims of our work; the provisioning of food for the present and into the future.

To do this, we must understand nature. This section is a glimpse into natural science as it pertains to the farm. In this way, we can work with the flow that surrounds us.

If the aim of society isn't the provisioning of health to the greater systems that support us, then we are destroying ourselves in the most primal way and fighting an unwinnable battle.

Earth's Spheres

How does nature express itself on, under and around your gardens, property and community? Having a good understanding of natural systems can help market gardeners maximize local resources, adapt to local climates, design for *ecosystem services* and avoid the costly mistake of working against the greater powers that be.

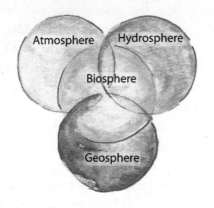

Earth's Spheres

Earth is composed of three primary spheres: air, water and earth, and biosphere, where life exists. The biosphere is the thin sphere of existence on our planet where air, water and earth support life.

- Atmosphere: Air
- Hydrosphere: Water
- Biosphere (ecosphere): Life
- Pedosphere: Soil
- Geosphere (lithosphere): Rock

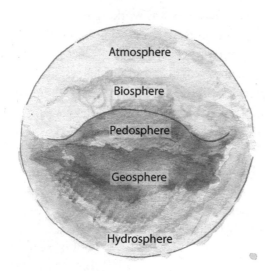

Earth's Spheres in Every Garden Bed

Earth's spheres can be found in every garden bed: water and air filling interstitial voids in the soil left by soil organisms, plants growing and animals roaming and breathing in the open air above the raw mineral material feeding and holding them.

Only in the last moment in history has the delusion arisen that people can flourish apart from the rest of the living world.

— E. O. Wilson

The Study of Earth

Natural Sciences

These spheres are ever interacting through flows of energy and matter. To best understand them, science has given us the following disciplines.

Climatology: The study of global, regional and localized movement of wind, water and weather patterns.

Hydrology: The study of water, its movement and reserves in the air, surface and below ground.

Geomorphology: The study of landforms, their creation, destruction and layout.

Biogeography: The study of the distribution of plants and animals through time and space.

North American Biomes

Biomes are regions designated by their climate (weather, temperature) and their plant and animal species. The same biome on different continents will not have the exact same complement of species, but they will have similar traits, having evolved in response to parallel environments. As such we find different succulent plants that can store water in their tissue across all desert regions and coniferous species of trees in northern forests.

This represents an opportunity for growers to look for production ideas on other continents within a similar biome. Understanding your biome is the first step in designing a ecosystem farm ecology you could mimic.

Geology: The study of Earth's rocks and minerals, their origins and cycles.

Pedology: The study of soil, its types, origins, layers and structures.

Edaphology: The study of soil's effect on living things.

Note: We will look at this most of all, not only because it is so directly important to farming, but also because it is a microcosm of all of Earth's spheres. The very life layer of earth with air, water, mineral and biological components dynamically at play.

Ecology: The branch of biology that deals with the relation of organisms to one another and to their physical surroundings.

Do You Know?

There are many kingdoms of life

Human Geography: The study of humans, our interactions, distribution, cultures and movement through space/time and our creation of place.

Do You Know?

Because these natural systems are always at play, it is important to place our farms (and their socio-economic context) *conceptually* within Earth's systems since they *actually* are.

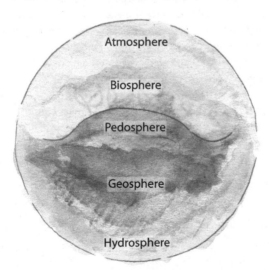

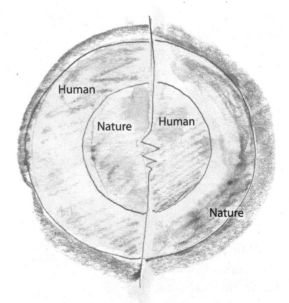

Humans and our socio-economics are within natural systems, not without.

Scientific Method

Here is a breakdown of the scientific method. Since it is a primary tool in science used for examining natural systems, I put it here for your interest and use.

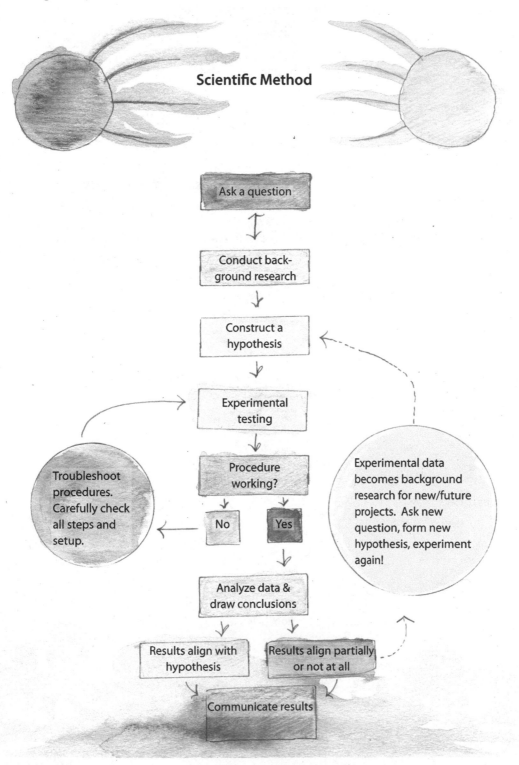

Scientific Method

Ask a question

Conduct background research

Construct a hypothesis

Experimental testing

Procedure working?

Troubleshoot procedures. Carefully check all steps and setup.

No   Yes

Experimental data becomes background research for new/future projects. Ask new question, form new hypothesis, experiment again!

Analyze data & draw conclusions

Results align with hypothesis

Results align partially or not at all

Communicate results

Levels of Organization

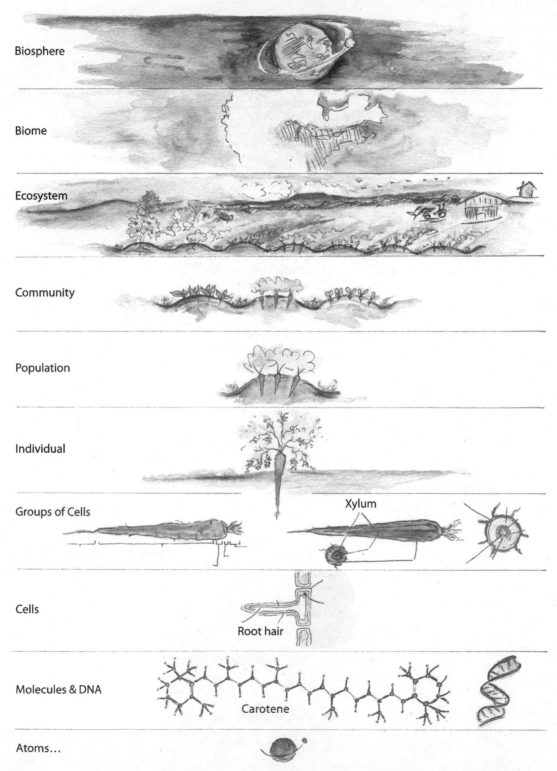

Organization of Life

Biosphere

Biome

Ecosystem

Community

Population

Individual

Groups of Cells

Xylum

Cells

Root hair

Molecules & DNA

Carotene

Atoms...

Scientific Classification

Classification of Life

Scientific classification, or taxonomy, helps make sense of the diversity of life.
It orders life in increasingly specific categories based on similarity of evolutionary characteristics.

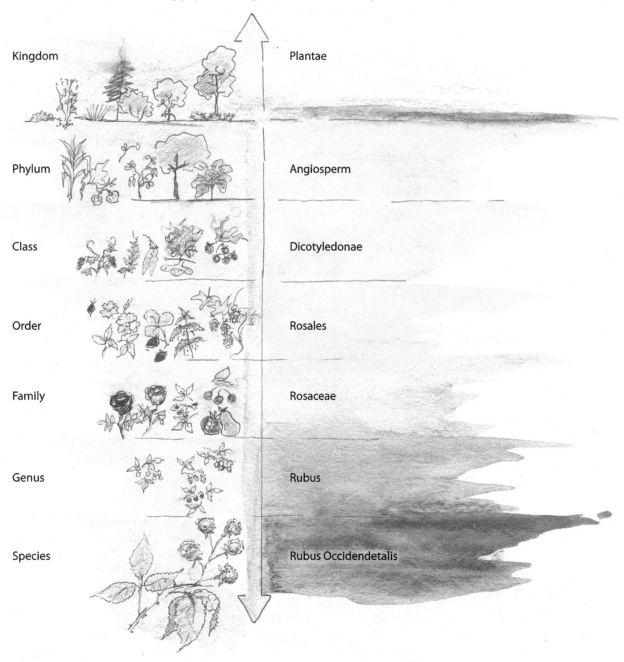

Kingdom	Plantae
Phylum	Angiosperm
Class	Dicotyledonae
Order	Rosales
Family	Rosaceae
Genus	Rubus
Species	Rubus Occidendetalis

To help remember these taxonomical groups from highest to lowest, try this mnemonic: King Peter's Class Ordered His Family's Genius into a New Species.

Black raspberries are one of our favourite crops and they are part of an important family, the rose family (Rosaceae) and the prevalent phylum of flowering plants (Angiosperm) in the mighty Kingdom of Plantae.

EXAMPLES OF NATURAL CYCLES AND SCIENCES

Farm Hydrological Cycle

The hydrological cycle is one of the most important natural cycles on a farm. Water is life!

Growing up in a desert, I know this only too well. The following pages show the full extent of the hydrological cycle on our farm.

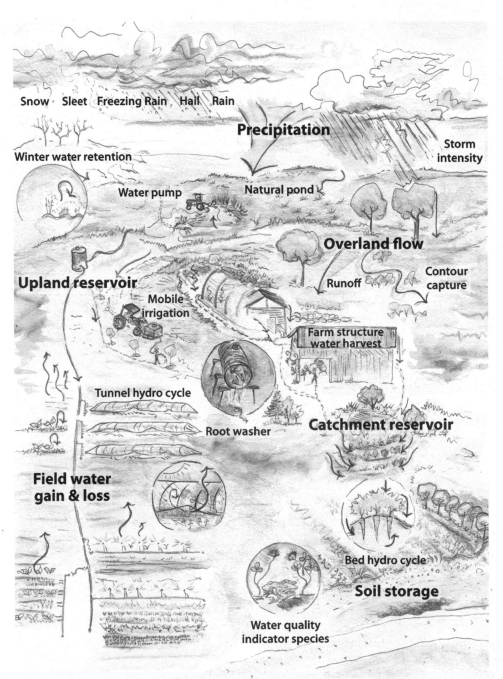

Snow   Sleet   Freezing Rain   Hail   Rain

Precipitation

Storm intensity

Winter water retention

Water pump   Natural pond

Overland flow

Contour capture

Runoff

Upland reservoir

Mobile irrigation

Farm structure water harvest

Tunnel hydro cycle

Root washer

Catchment reservoir

Field water gain & loss

Bed hydro cycle

Soil storage

Water quality indicator species

The hydrological cycle is a system of flows, stores and interactions. It is one cycle, but we can look at it from both natural and human sides.

The flows:

- Evaporation
- Transpiration
- Condensation
- Precipitation
- Overland flow
- Interception
- Percolation
- Water table flow
- Aquifer storage

Some examples of farm water cycle:

Capture

- Rooftop
- Ponds
- Contours
- Swales and berms

Infiltrate

- Biopores
- Soil structure
- Root space
- Mulched surfaces

Store

- Ponds
- Swales
- Gravity tanks
- Ice in root cellar
- Septic

Move

- Hose
- Pallet tanks
- Contours
- Berms

Release

- Earthworks
- Drip
- Sprinkler
- Spray gun
- Root washer
- Ice melt
- Weeping bed
- Greywater
- Water trough

Conserve

- Reduce evaporation
- Make efficient use
- Water-holding capacity of soils
- Xeriscaping
- Improved plant use

- Monitor leaks
- Intercropping, cover cropping, constant soil cover
- Mulching
- Microclimate cycling
- Organic soil

Concern for Water

Water is a valuable resource for agriculture that is too often misused. When combined with unfamiliar precipitation trends of global climate change, farm hydrology is increasingly under scrutiny for improved management of its collection, storage, transport and use.

We work to conserve water through many approaches. We harvest and store water in ponds and storage tanks and use water conserving irrigation systems and strategies. But

Water has three states: liquid, solid and vapour. There is latent heat lost or absorbed with state transfer. This is how our passive cold storage works. We fill it each winter with 1,000 cubic feet of ice; as it melts our cellar temperature drops.

Amphibians are an indicator species of a healthy environment, especially water. Their permeable skin is susceptible to lack of this resource and to water pollutants.

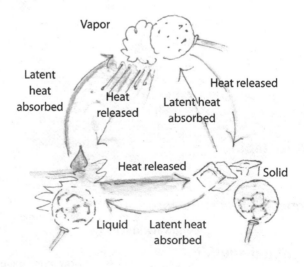

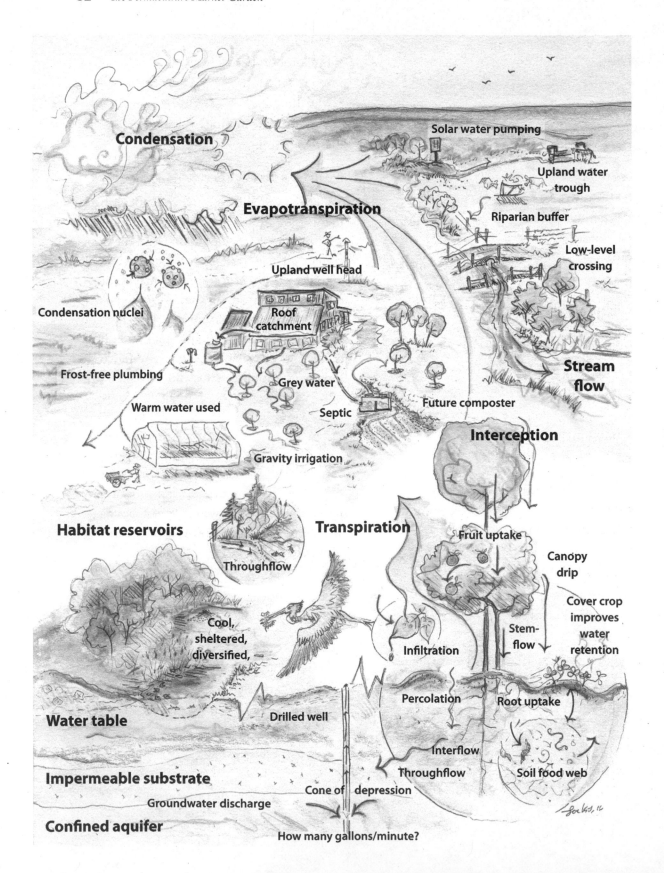

Condensation

Solar water pumping

Upland water trough

Evapotranspiration

Riparian buffer

Condensation nuclei

Upland well head

Low-level crossing

Roof catchment

Frost-free plumbing

Stream flow

Grey water

Warm water used

Septic

Future composter

Gravity irrigation

Interception

Habitat reservoirs

Transpiration

Fruit uptake

Canopy drip

Throughflow

Cover crop improves water retention

Cool, sheltered, diversified,

Infiltration

Stem-flow

Water table

Drilled well

Percolation

Root uptake

Impermeable substrate

Interflow

Throughflow

Soil food web

Groundwater discharge

Cone of depression

Confined aquifer

How many gallons/minute?

SLOW IT, SPREAD IT, SINK IT, SAVE IT

also of concern is building the soil's capacity for absorbing, holding and releasing water for crop growth, improving garden micro-climates for localized water regulation, and crop selection and seasonal rotation for water efficiency.

Market gardeners grow many tender annuals throughout the summer when it is unnatural for young succulent lettuce, spinach and other crops to even grow. Market demand year round for these products is immense. Yet, it is nearly as unnatural for young spring greens to grow in July as it is for them to do so in January. How can we better nurture cool and moist conditions for spinach in summer without excessive sprinkler waste? Overwatering with sprinklers can result in excess water loss to evaporation and even compact the top inch of soil from the incessant pounding of rain, resulting in runoff over an impenetrable soil crust.

There is not always enough rain in July when our storage carrots are seeded to maintain ideal soil moisture. Without rain, our clay soil can crack, exposing the young roots to dry air. Our storage carrots require moisture to be maintained in the uppermost garden soil horizon for 2 to 4 weeks of germination and young growth before the carrot can reach deeper stores of water on its own.

We need to improve the soil's capacity to hold and regulate water. This can be achieved by enhancing soil organic matter, using surface mulches and building a loose soil

GREEN THUMB TECHNIQUE

Planting shade trees around ponds, along streams and within your fields cools the environment, buffers against nutrient losses and helps conserve water. It also creates habitats for beneficial organisms, like amphibians and dragonflies.

GREEN THUMB TECHNIQUE

Garlic

Many diseases and pests react to water availability. Make observations in wet and dry years to inform your farm management. For instance, *bulb and stem nematodes* travel in soil water.

If your garlic is infected, years of abundant rain will exacerbate the problem, revealing yellowing leaves and bloated lower stems. Do not save as seed!

Persian Star

Grow demand for seasonal eating since seasonal crops require less energy and management to produce.

structure through the working of soil microbiology and the decay of their bodies and plant residues into biopores. This improved soil can better capture water from heavy rain events and hold it like a sponge for slow release to plant roots and local aquifers. High organic matter in the garden bed's O horizon (see page 43) helps maintain soil moisture in the critical 3-inch layer needed for seed germination and young growth between rains.

In addition, we should consider how we could modify the garden ecosystem against desiccation and evaporation by wind and sun through garden microclimate modification and by integrating layers of diversified crops. Integrating cover crops with vegetable crops can provide shade, windbreaks and future organic matter. Examples include using rye cover crops alternated with late spring crops and buckwheat with summer salad crops.

Trees integrated into the garden landscape have a similar effect, breaking wind, providing shade and regulating the local water cycle. A layered garden of trees, shrubs, cover crops and vegetable slows rainfall, intercepting rain at different heights and releasing it slowly through leaf drip and trunk flow to the soil surface. Organic matter and mulch on the beds also slow the rain, allowing soil water to percolate into the soil and prevent soil saturation, which can result in overland flow and erosion. When soil is saturated, then rain cannot infiltrate into its pore spaces.

Finally, we should balance our production of tender crops in the summer months with crops that are better suited: more heat- and drought-tolerant, crops with deep taproots or ones that lend themselves to mulched production systems. For instance, winter squash have already sent their roots deep and have shaded over the soil surface to conserve moisture. It just really isn't that sustainable to mass grow spinach and radish through a July drought! But if you are going to, consider how to model your garden's soil and diversity for improved water regulation

Geological Cycle: Ottawa Valley Stratigraphy

Stratigraphy is the study of rock and sediment strata and their layering over geologic time. This gives us perspective into our relatively short habitation of Earth. More to the point, this reveals the origins of soils we currently farm and indicates opportunities for locally available mineral resources, including our very own soils and subsoil. What follows is a breakdown of the geological activity in my local region, with some examples of places and occurrences for reference.

Note: These geological events are ordered oldest to most recent.

1. **Precambrian Canadian Shield (Grenville Province) is divided along tectonic terrains (circa 1250-980 Ma).**

   a. Metamorphosis and deformation of sedimentary rocks (gneiss and marbles)

   b. Intrusive plutonic rocks like granite, syenite, gabbro and diorite

   c. All of these rocks are part of an ancient Proterozoic mountain belt

2. **Unconformity from erosion and subaerial weathering of 100s of millions of years.**

   a. Strata from this time period is absent

3. **Paleozoic sedimentary rocks (ca. 523 Ma to 438 Ma)**

   a. Geologic strata of shale, dolostone, limestone and sandstone are found locally. Our bush lot has wonderful limestone sink holes. The rocks were deposited in terrestrial and marine conditions around an ancient epeiric sea that once lapped the margins of the aging Precambrian Shield

There are three types of rocks: sedimentary, igneous and metamorphic.

1) Sedimentary rocks are created from compression of loose sediment into rock.

2) Igneous are made either within the guts of the Earth (intrusive, like plutonic) or by exiting and solidifying in open air (extrusive).

3) Metamorphic rocks are made through the high pressure and temperature and partial melting of the rock, resulting in foliation, when like crystals align (gneiss) and new elemental combinations occur (marble from limestone).

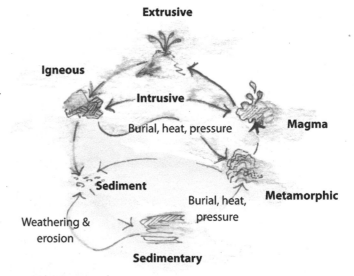

The rock cycle explains the origins of our rocks, which become the parent materials for our soils.

4. ***Paleozoic faulting causes* west-east and northwest-southeast faulting, forming the Ottawa-Bonnechere Graben, our valley.**

5. **Pleistocene glaciation results in unconsolidated sediments (2.58 Ma to 12,000 years ago).**

   a. The oldest deposits are unconsolidated clay-rich glacial till: moraines and eskers, etc., from the Wisconsin ice sheets advance and retreat across much of northern North America

   b. A series of glaciofluvial and glaciolacustrine events following ice retreat produced much of the stratified sand and coarse gravel that sheet the Ottawa environs.

   c. Finer sediment, like the Leda clay in our northern farm field (70 meters thick), was deposited during a period of sea incursion into the depressed post-glacial landscape.

   d. This new inland Champlain Sea accumulated fine sediments in its depths and left these deep accumulations and sedimentary beaches when it retreated to the east after land's isostatic rebound.

   e. The last stage in a dynamic and drawn-out forming of our geology is the release of immense volumes of ice water with the continued retreat of the glaciers to the north.

   f. This water drained through the Ottawa valley and Gatineau River valley helping form the landscape and reform it again and again — eroding and depositing older till and marine deposits over time.

   g. This also formed many of the terraces and bogs during channel reestablishment that we can find (such as Mer Bleue bog).

   h. This eventually stabilized into the current river channel systems we know today, and the agricultural soils and solum we farm developed from this immense diversity of parent material.

Civilization exists by geological consent, subject to change without notice.

—Will Durant

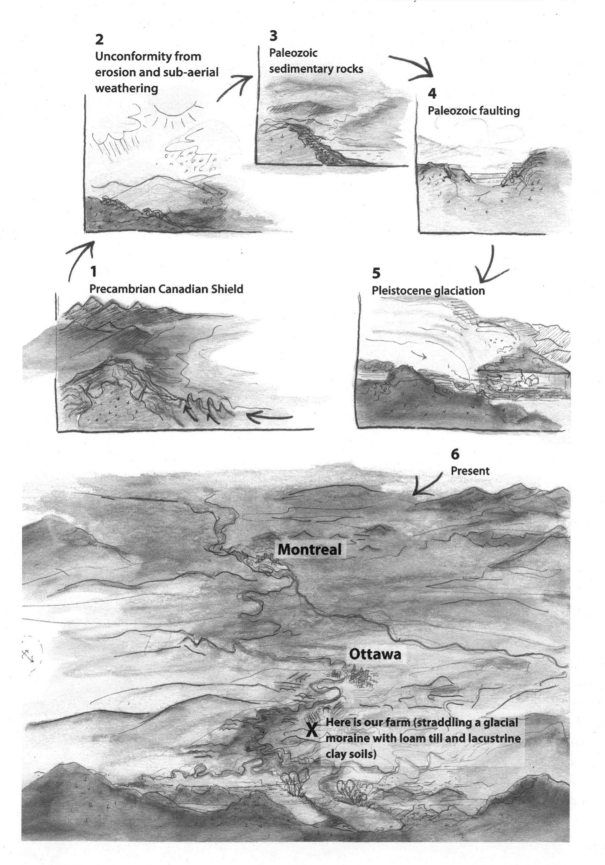

2
Unconformity from erosion and sub-aerial weathering

3
Paleozoic sedimentary rocks

4
Paleozoic faulting

1
Precambrian Canadian Shield

5
Pleistocene glaciation

6
Present

Montreal

Ottawa

X **Here is our farm (straddling a glacial moraine with loam till and lacustrine clay soils)**

A CLOSER LOOK AT SOIL

Pedology

Pedology is the study of soil:

- How soil is made (pedogenesis)
- Soil morphology
- Soil classification

The **pedosphere** is the outer most layer of our planet where soil processes help us grow our crops. It's the supportive life layer where all of Earth's spheres interact constantly.

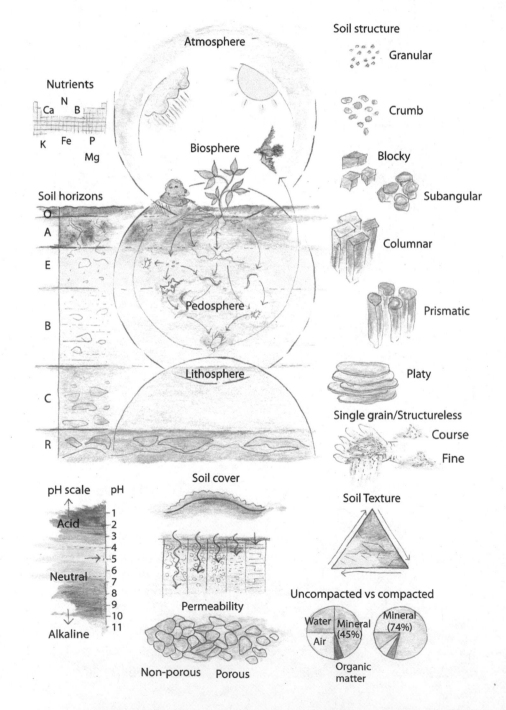

Do You Know?

Many vegetables prefer a more alkaline soil: cabbage, thyme and cucumber; whereas rhubarb, sorrel and sweet potato like soil to be very acid. Most garden soils range from pH 5 to 8.

Pedogenesis is the process of soil forming. It is important to understand soil formation because it teaches about properties of the soils we farm and can influence design for regeneration of soil benefits: nutrient cycling, loose structure and habitat for improved production.

Pedogenesis

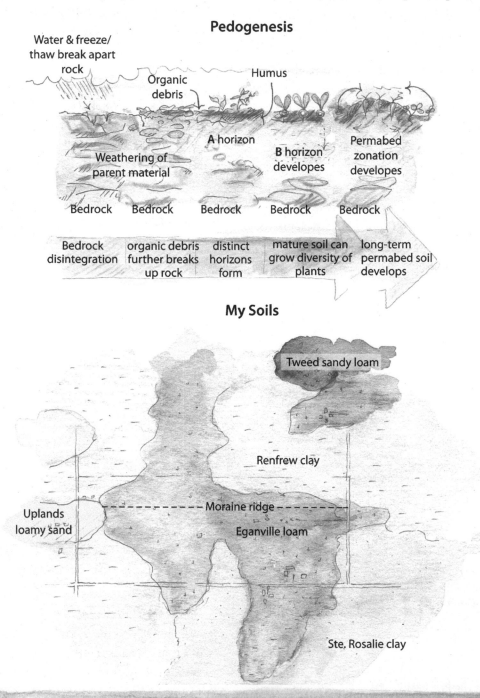

Water & freeze/thaw break apart rock

Organic debris

Humus

Weathering of parent material

A horizon

B horizon developes

Permabed zonation developes

Bedrock   Bedrock   Bedrock   Bedrock   Bedrock

Bedrock disintegration	organic debris further breaks up rock	distinct horizons form	mature soil can grow diversity of plants	long-term permabed soil develops

My Soils

Tweed sandy loam

Renfrew clay

Uplands loamy sand

Moraine ridge

Eganville loam

Ste. Rosalie clay

Parent material is the original material that was deposited or uplifted from which a soil has formed. On our farm this is glacial till and lacustrine clay.

Soil Formation

This is important to understand because it will help us design systems for increasing soil health on the farm. There are five primary factors involved.

- **Parent Material** is the raw mineral material that makes up the base component for any future soils. There are different types of parent material, and most are geologic in origin (in our case: glacial till and inland sea clay), with an exception for peat-based soils.

    The parent material will greatly determine the character of a soil: the elements available for soil organisms to work with in forming the new soil, the way water is held or moved, the structure of the soil. This will affect how we choose to farm a soil. Sandy, clay or stony, this has a big impact on the equipment, crops and management practices for market gardeners.

In stony soil, stones just keep coming up as you work it over the years. Maybe we should leave them down and build layers on top. We now grow only in raised beds, and we leave the bones of the Earth where they belong, supporting our bed.

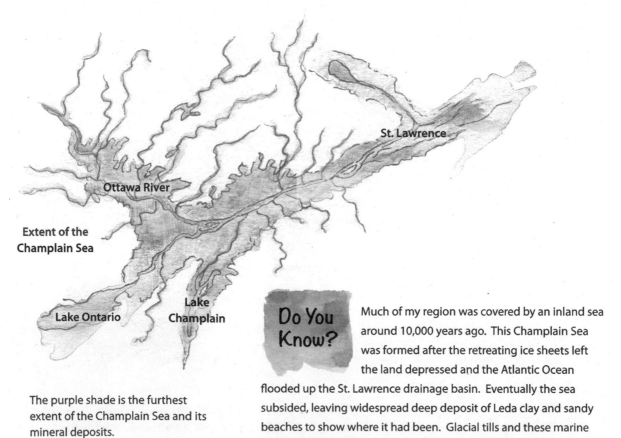

The purple shade is the furthest extent of the Champlain Sea and its mineral deposits.

Do You Know? Much of my region was covered by an inland sea around 10,000 years ago. This Champlain Sea was formed after the retreating ice sheets left the land depressed and the Atlantic Ocean flooded up the St. Lawrence drainage basin. Eventually the sea subsided, leaving widespread deep deposit of Leda clay and sandy beaches to show where it had been. Glacial tills and these marine deposits constitute the majority of parent material from which regional soils developed.

- **Climate** and its weather systems break up, erode, disperse and settle mineral and organic materials contributing to soil formation. Physical and chemical weathering are major contributors to soil formation. Climate also determines the rate of chemical weathering (enhancing rock degradation with increased warmth and precipitation) and determines which organisms are present to interact within the soil.

- **Organisms** that live in and on the soil have a great effect on soil formation. They contribute organic matter through leaf fall, feces and decomposed organisms. They can alter the pH (conifers have acidic leaf litter), and they can help generate rich organic humus layers.

- **Topography** will increase or decrease the effect of wind, water and gravity at work on the parent material in producing soil

- **Time** is the magic ingredient.

Mottling shows up in soil as the reddish brown spots from the oxidation of soil elements and can indicate a fluctuating water table. Gleying, or gray or bluish colors are formed from waterlogged soils.

Ultimately, the only wealth that can sustain any community, economy or nation is derived from the photosynthetic process — green plants growing on regenerating soil.

— Allan Savory

GREEN THUMB TECHNIQUE

Cover Crops

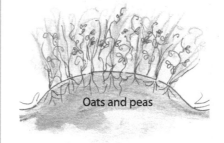

Oats and peas

Oilseed radish

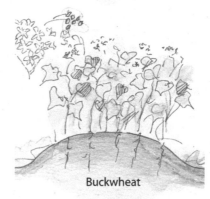

Buckwheat

Cover crops can benefit the farm in many ways. Legumes like peas and clovers fix nitrogen, tap-rooted red clovers and oilseed radish loosen the soil and accumulate deep stores of nutrients, and broad-leaved buckwheat quickly shades and outcompetes weeds. All provide organic matter and protect and stabilize the soil.

Soil Structure

The arrangement of soil aggregates into different forms gives a soil its structure. The natural processes that aid in forming aggregates are:

- Wetting and drying
- Freezing and thawing
- Soil life activity that aids in the decay of organic matter and movement of earth
- Activity of roots and soil animals
- Absorbed cations

The actions of wetting/drying and freezing/thawing, as well as root and animal activity, push particles back and forth to form aggregates. Decaying plant residues and microbial by-products coat soil particles and bind particles into aggregates. *Absorbed cations help form aggregates whenever a cation is bonded to two or more particles.* A healthy soil has higher cation exchange capacity, increasing the ability of plant to access nutrients.

Principle

Cover crops have many benefits, such as promoting soil-building activities by maximizing food sources for soil life and protecting them against the elements. Every season has its cover crop. Understanding cover crops, their growth rate, frost sensitivity, form and function can help you place them into your crop rotation to avoid periods of fragile bare soil.

Do You Know? The difference between physical and chemical weathering: physical weathering is the breaking up of parent material through freeze/thaw action or plant-root penetration of cracks. On the other hand, water can react with elements through hydrolysis (chemical weathering).

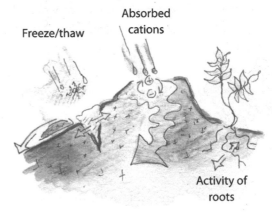

Freeze/thaw

Absorbed cations

Activity of roots

Summer | Fall

Winter | Spring

GREEN THUMB TECHNIQUE

Use a soil auger to analyze all of your soil's horizons. Recognizing mottles and gleying can help you determine a soil's drainage and moisture regime even in a dry season or late summer when water may not be as present in the soil profile.

Soil Horizons

As soil forms, the profile differentiates to create horizons. **Soil horizons** are parallel layers in a soil profile that differ significantly from those above and below. These are identified through attributes like color (brown, umber, grey), texture (sand, silt, clay) and specific features (like gleying, mottling).

Basic Horizons

L, F, H — Horizon prefixes. There is an increasing material decomposition and a decreasing visibility of organic leaves and twigs from L (leaf litter) to F (folic) and finally to H (humic). Typical of more upland forested soils.

O — An organic soil horizon composed from decomposed bog vegetation, like peat. Typical of lower, wetland sites.

A — A mineral horizon near the surface mostly defined by its accumulation of organic materials from above and its loss of materials to lower horizons. It can be subdivided due to accumulation of organic matter (Ah) and loss of clays, iron and aluminum from eluviation (Ae).

Soil texture is determined by ratio of different particle sizes in a given soil. The texture triangle shows the texture relationships between sand, silt and clay. Pouring some of your soil into a jar of water and letting it settle can often give you a good idea of the ratio between these soil particle types.

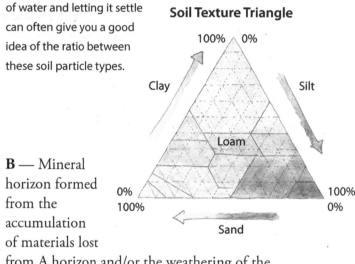

B — Mineral horizon formed from the accumulation of materials lost from A horizon and/or the weathering of the parent material below.

C — This horizon is characterized by minimal influence from pedogenic processes. Some variations occur through: accumulation of Ca and Mg or salts or presence of mottling and gleying due to water saturation.

R — Consolidated bedrock.

W — Water layer.

Our Soil: A profile of Eganville Loam

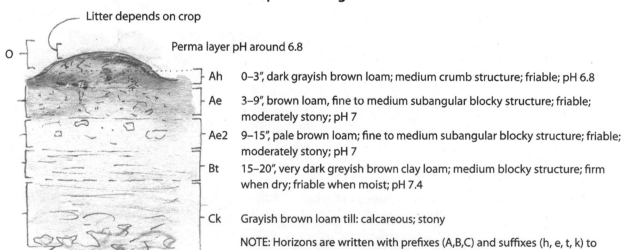

Litter depends on crop

Perma layer pH around 6.8

O

Ah — 0–3″, dark grayish brown loam; medium crumb structure; friable; pH 6.8

Ae — 3–9″, brown loam, fine to medium subangular blocky structure; friable; moderately stony; pH 7

Ae2 — 9–15″, pale brown loam; fine to medium subangular blocky structure; friable; moderately stony; pH 7

Bt — 15–20″, very dark greyish brown clay loam; medium blocky structure; firm when dry; friable when moist; pH 7.4

Ck — Grayish brown loam till: calcareous; stony

NOTE: Horizons are written with prefixes (A,B,C) and suffixes (h, e, t, k) to better specify attributes found in soils in short form.

Soil Nutrients

It is important to test your soils for their nutrient content. Not only should we understand the availability of the macronutrients (nitrogen, potassium and phosphorus), but also a slew of micronutrients. The following list exemplifies the 16 essential nutrients for plant health.

Periodic Table

Understand what is most available to your plants

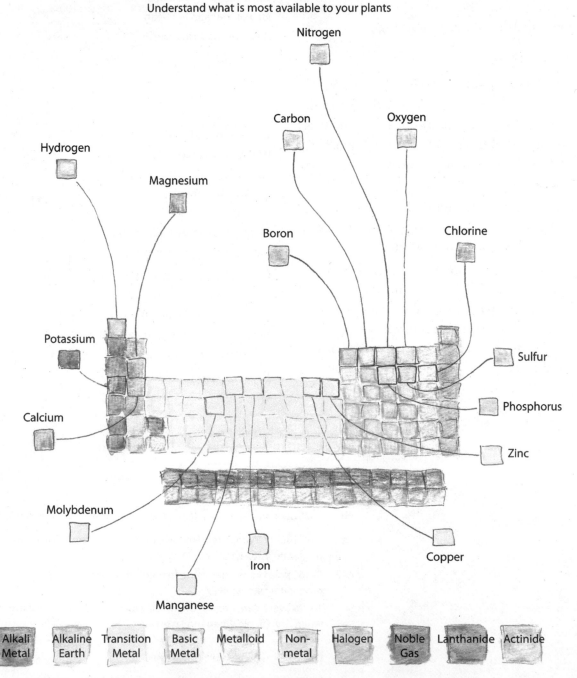

Fix and cycle nitrogen, for it composes 78% of our atmosphere.

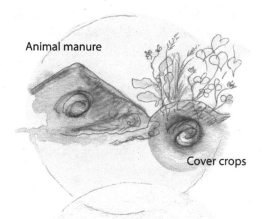

Animal manure

Cover crops

Micronutrients

Liquid fertility

Crop-cover crops

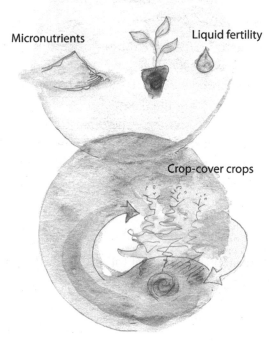

It is important to understand sources of nutrients and brainstorm alternatives to bagged commercial products. Local sources include:

- Ocean, pond and river nutrients
- Quarry tailings
- Animal manures
- Farm construction topsoil removal
- Sequestration and fixation by cover crops and perennials

When starting a market garden, it is important to jump-start your soil. See the sidebar for our nutrient routine. We boost our soils with Spanish River Carbonatite, a great mineral resource. Thanks to Marc Beauchamp for introducing us to this natural blend of over 40 micronutrients. Remember that plant nutrition depends on more than just nutrient presence in a soil. Complex soil interactions and health is key! Water, organic matter, mycorrhizal fungi and other soil life help gather, store and release nutrients to plants.

16 Essential Nutrients

There are many elements on our planet. Just like humans are more functional when well fed with nutrient-dense foods, plants depend on adequate supply and use of nutrients to grow and maintain healthy structure, tissues and functions. *We cannot expect plants to grow and provide us edible leaves, roots and fruits*

GREEN THUMB TECHNIQUE

When opening new land for gardening, it is important to boost fertility, enhance the soil's capacity to cycle nutrients and design for increased inter-farm waste dynamics for macro-fertility cycling.

and put up defense against insects and disease, if they don't have the building blocks and foods for their forms and functions.

THE THREE AMIGOS

Carbon (C) is acquired from our atmospheric reserves of CO_2 through photosynthesis and used to produce glucose for plant use and releasing oxygen.

Hydrogen (H) helps form glucose, which is essential for producing energy and synthesizing amino acids and nucleotides.

Oxygen (O_2) is derived from air and water and is also a building block for glucose and a byproduct of its creation. Oxygen is crucial for plant uptake of nutrients in the rhizosphere and for nutrient transfer across cellular walls. Macro and micro pore space in the soil is needed to maintain adequate oxygen levels. A compacted or flooded soil will starve a plant.

MACRONUTRIENTS

Nitrogen (N) is critical for plant growth and leafing; its role in green growth can be seen

GREEN THUMB TECHNIQUE

You can fine-tune the growth and production of a plant through well-timed nutrient management. Tomatoes need more nitrogen in spring for leaf and stem growth and less later on when fruits are ripening. We can balance their needs through light compost and clover cover crops to give a boost of nitrogen in spring, but less in fall.

in tomato stalks as they thicken and thin with nitrogen uptake. It is used in fairly large quantities and considered a primary nutrient. It is involved in photosynthesis, formation of amino acids, cellular division and other essential processes.

Phosphorus (P) is found in rocks and in water in a soluble form. Phosphorus is readily eroded and is a chief cause of water pollution in agricultural areas. It is a limiting nutrient crucial for early root formation, cell division and growth, fruit and seed ripening and overall plant maturation. Phosphorus is essential because it transforms energy from the sun and other chemicals into food for the plant.

- Low phosphorus levels result in stunted growth and dark green and purplish colors in leaves; this is caused by buildup of sugars and development of anthocyanins.

Potassium (K), derived from rocks, is important right from the get-go; absorbed in quantity through the roots, it helps plants

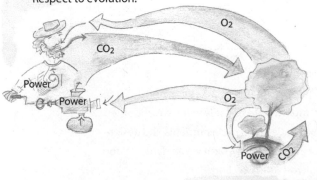

There is a long-standing relationship between oxygen-breathing organisms and carbon-sequestering plants. Respect to evolution!

grow, managing water use efficiency, protein production, fruit and seed formation and ripening. It is very important in the market garden as it contributes to winter hardiness and disease resistance.

SECONDARY NUTRIENTS

Calcium (Ca) is another mover and shaker. It is critical for the mobility of nutrients taken up through the roots and responsible for cell division and increase cell wall stability, holding the plant up and keeping its cellular networks open.

Magnesium (Mg) is important in the manufacture of chlorophyll, used for photosynthesis. It improves the use and movement of phosphorus and iron.

Sulfur (S) is used in nodule formation for nitrogen fixation by legumes. It is also used for chlorophyll production and is essential to amino acid, enzyme and vitamin formation.

MICRONUTRIENTS

Iron (Fe) is critical, although only relatively small amounts are needed, for respiration and photosynthesis, contributing to chlorophyll formation and movement of oxygen.

Manganese (Mn) contributes to carbon dioxide uptake in photosynthesis, chlorophyll synthesis and enzyme system functions.

Copper (Cu) is important for chlorophyll production and is involved in the reproductive growth. It affects fruit and seed formation, coloring and flavoring.

Zinc (Zn) is crucial for plant growth through the formation of auxins. Interestingly, insufficient zinc results in shorter distance between plant growth nodes. It is needed for

production of chlorophyll, carbohydrates and starches.

Boron (B) is critical to plant cellular development, resulting in stronger cell walls and the formation of fine structures like root tips, buds, seeds, pollen tubes and plant storage tissue. Boron deficiency contributes to scarring in beets and hollow stems in broccoli.

Molybdenum (Mo) is involved in the conversion of nitrates and phosphates into useable plant nutrients through enzymatic systems.

Chlorine (Cl) is one of the most plentiful elements. Chlorides are salts derived from chlorine and another element. They are useful for enzyme activity, water regulation and nutrient movement. Chloride deficiency in wheat shows up as a yellowish mottle on leaves.

Principle

Holistic Plant Nutrition

Prioritize feeding the soil and building its nutrient storage and cycling capacity so plants can fend for themselves.

GREEN THUMB TECHNIQUE

Soil tests should be conducted on a regular basis for benchmarks in soil improvement and to help fine-tune plant health. Soil testing at the same time of year improves accuracy of year-to-year comparisons.

Nutrient Cycles

There are many nutrient cycles. We'll look at phosphorus and nitrogen.

TYPES OF PHOSPHORUS

- Particulate phosphorus: bound up in sediment or organic matter, longer lasting in environment
- Soluble phosphorus: in a solution, much smaller and more available to plants

ABOUT PHOSPHORUS

- This is the second most limiting nutrient to plant health after the ubiquitous nitrogen. It makes up about .2% of plant dry matter
- Much phosphorus can be found in mineral soils, especially in clay. However, this soil phosphorus is often unavailable to plants. This nutrient is locked up in forms that cannot be assimilated by plants, and the

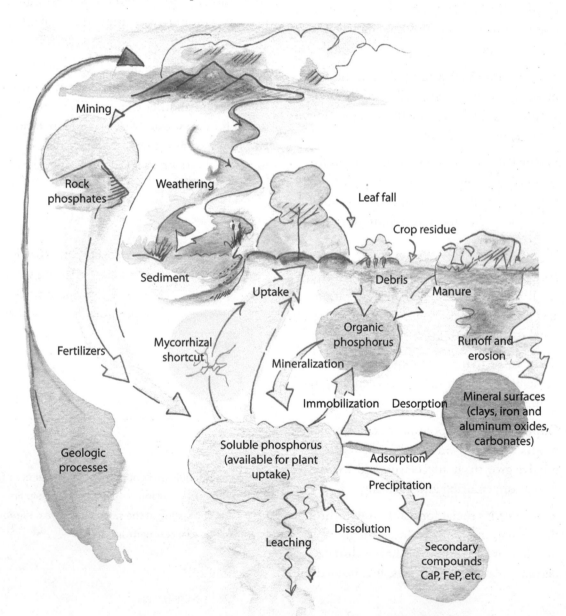

process of making it available is often too slow for fast-paced agriculture. Modern agriculture is dependent on fine ground phosphate fertilizers that run off plowed fields easily, contributing to the *eutrophication* of our lakes and rivers and the draining of the farmer's pocketbook.

- Dissimilar to nitrogen, phosphorus is bound to soil particles. This can be found in secondary compounds with iron, aluminum, calcium and magnesium. Soluble phosphorus can *precipitate* as these compounds and return to soil solution through *dissolution*. Phosphorus is often bound in clays as iron and aluminum oxides and carbonates through adsorption and can be returned to soil solution through desorption.

- Soils low in aluminum and iron (very sandy or high organic matters) are less able to bind phosphorus and likely to have greater losses from leaching.

MINERALIZATION

- Soil microorganisms break down phosphorus found in crop residues and animal manure, releasing readily available phosphorus ions through their decomposition process. This soluble phosphorus is available to plants.

IMMOBILIZATION

- Process of soluble inorganic phosphorus converted to organic phosphorus. This helps hold phosphorus in reserve for future uptake.

ASSIMILATION

- Plants can assimilate phosphorus when it is in a soluble form. Phosphorus ions that are not soon taken up by plants can be adsorbed onto particles of soil. Healthy soil life can help continual cycling of phosphorus, and mycorrhizal fungi, in particular, can aid crops to quickly assimilate available phosphorus.

- Plants and fungi take up phosphorus as inorganic phosphates, phosphorus bound to oxygen and hydrogen. When plant material dies, it returns to the soil as organic phosphorus that can either be mineralized and made available again or immobilized for organic storage.

- Arbuscular mycorrhizal fungi (see Life Within the Soil, (page x) increase phosphate uptake and increase its availability in the soi.

- Without doubt, presence of mycorrhizal fungi in soil quickens translocation of phosphorus from soil through crop cellular walls.

 Do You Know?

We should think of soil life as tools like any other: helpful when we understand how to use them and requiring our maintenance for better performance.

Tools come in many forms: mycorrhizae, stirrup hoe, power harrow.

 Soluble nutrients are available for immediate plant uptake from the soil solution, but they can also be lost through leaching and erosion. A complex food web allows nutrients to be mineralized/available and immobilized/stored.

On-farm Nitrogen Cycle

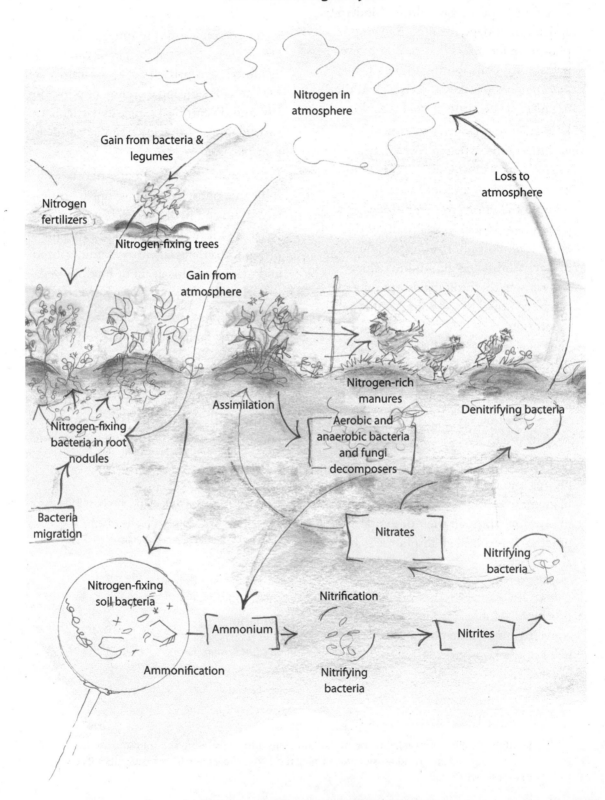

The Life Within the Soil

Plants are in constant interaction with soil life, including bacteria, fungi, protozoa, nematodes, arthropods and earthworms. Plant growth without soil life is like playing without a full hand of cards, not worth gambling on! We must understand the soil organisms involved so we can value and use their services.

There are many players in the soil food web. They are in a constant dance of nutrient acquisition, release and exchange. They eat, excrete, move and die, building soil spaces, decomposing organic matter and cycling nutrients. They make nutrients available for plants and form symbiotic relationships with our crops to improve resource provisioning.

"I'd gamble my entire crop, but I'm missing the Ace of Mycorrhizal."

1. Bacteria

The majority of soil bacteria are decomposers, consuming *root exudates* and other organic matter. Bacteria immobilize the nutrients held in soil organic matter, making them available for other soil organisms.

Nitrogen-fixing bacteria, called *rhizobia*, are an awesome friend to the farmer. They can acquire nitrogen directly from the atmosphere, turning it into ammonium in the soil. Legumes, such as peas, beans or clovers, will sent out flavonoids to alert the bacteria and form specialized root nodules for the bacteria to set up shop for the *fixation* of hundreds of pounds of nitrogen per acre.

Do You Know?

Chemoautotrophs, another bacterium, help degrade pollutants and also cycle nitrogen. This is especially important in areas where agriculture is replacing industrial yards or where many years of conventional agriculture have left lasting pollutant residues.

2. Fungi

Fungi help stabilize our soils, regulate soil water, increase plant nutrient uptake and suppress disease.

Saprophytic fungi serve as decomposers; they are masters of turning our dead waste into food for soil life, helping cycle nutrients back up the food chain to our crops and animals. As they grow, building fungal mass, they convert organic matter into organic acids and release CO_2. This works

Root exudates are sweet juices excreted by plant roots to feed soil life in exchange for services like nitrogen fixation by bacteria and nutrient exchange with fungi.

Permabed Soil Food Web

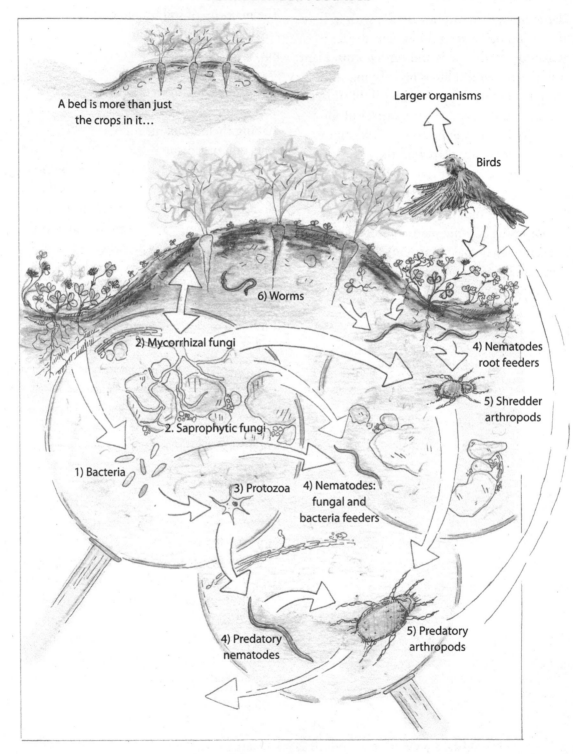

A bed is more than just the crops in it…

Larger organisms

Birds

6) Worms

2) Mycorrhizal fungi

2. Saprophytic fungi

4) Nematodes root feeders

5) Shredder arthropods

1) Bacteria

3) Protozoa

4) Nematodes: fungal and bacteria feeders

4) Predatory nematodes

5) Predatory arthropods

to *immobilize* and store nutrients for future release. The production of *humic-acid-rich organic matter*, which is much less readily degraded, can help stabilize the soil for hundreds of years.

Mycorrhizal fungi form mutual relationships with plants. There are two types:

- **Endomycorrhizae** commonly associate with crops, vegetables and shrubs.
- **Ectomycorrhizae** associate with many trees.

About 80% of terrestrial plants form relationships with endomycorrhizae, and many others form relationships with ectomycorrhizae.

Mycorrhizal fungi have an important role *in solubilizing phosphorus* and making it available for plant uptake. These fungi create exchange interfaces around or within plant roots and provide phosphorus, nitrogen, micronutrients and possibly even water to plants and receive sugars.

3. Protozoa (protists)

These *unicellular eukaryotic* organisms, such as flagellates and amoebas, are simple, little and mobile, consuming high amounts of nitrogen-rich bacteria and releasing ammonium. This ammonium is mineralized for plant and soil organism use. Protozoa are prey for larger nematodes and other micro-invertebrates, cycling their nutrient rich bodies up the food chain.

4. Nematodes

Nematodes feed on roots, bacteria and fungi and play important roles in stimulating plant growth and microorganism populations through grazing. *Grazing* has an effect similar to cows rotated on grassland; microorganism populations return with more vigor like grass regeneration. Ammonium is released in the process, which feeds the rejuvenation. Nematodes also act as dispersers of bacteria and fungi through carrying live and dormant microbes in their gut and on their body.

Do You Know? Predatory nematodes are seemingly more sensitive to soil disturbance and can be used as an *indicator* species of healthier soil.

5. Arthropods

These ubiquitous creatures include many garden invertebrates with *exoskeletons*, including insects (beneficial Assassin bugs, Braconid wasps and ladybirds, etc.), millipedes and centipedes (Myriapods), spiders (arachnid) and sow bugs (a terrestrial crustacean) responsible for shredding of organic matter.

Although some consume and shred roots and organic matter, helping to break it down for further assimilation of nutrients, most are helping cycle nutrients through predation. Arthropods are the construction crew of soil by opening soil spaces that are invaluable for water, air and nutrient movement.

Specialist predators can be a powerful ally against specific pests. It is the generalist predatory arthropods who serve as microjuggernauts, a force for many pests to reckon with.

Do You Know?

Parasitic wasps can help biologically control pests like leek moth, tomato hornworm and cabbage looper. Release of these beneficials should be timed with pest population cycles.

6. Earthworms

Earthworms' contribution to soil structure and aggregate improvement is becoming increasingly important as farmers realize that many of the nutrients we apply to soils are lost with poor *water-holding capacity*.

Earthworms are shredders of organic matter, movers of soil, buriers of organic material and tunnel miners. If arthropods are the crew, worms are the foremen, doing the heavy work so smaller organisms can extrapolate on soil spaces and digest material for further nutrient release. Earthworm tunnels are coated in nutrients, easily accessible to plant roots and fungal hyphae. The *castings* left by their work have been shown to be a super fertilizer, much richer in soluble nutrients and soil microorganisms than material consumed by the earthworm.

Protect the soil and improve habitat for soil life. Without fail I find earthworms near cover crops in spring and not in bare soil. They are immediately available to improve my spring beds!

Worm castings are earthworm excrement, which is an excellent microfertilizer. You can buy this as a product and/or encourage earthworms in your garden with cover crops, manure and mulch. This is also a great amendment for greenhouse starts. I call the total excrement of soil life micromanure!

Soil Life per Teaspoon

Note: Earthworms and arthropods are per square foot

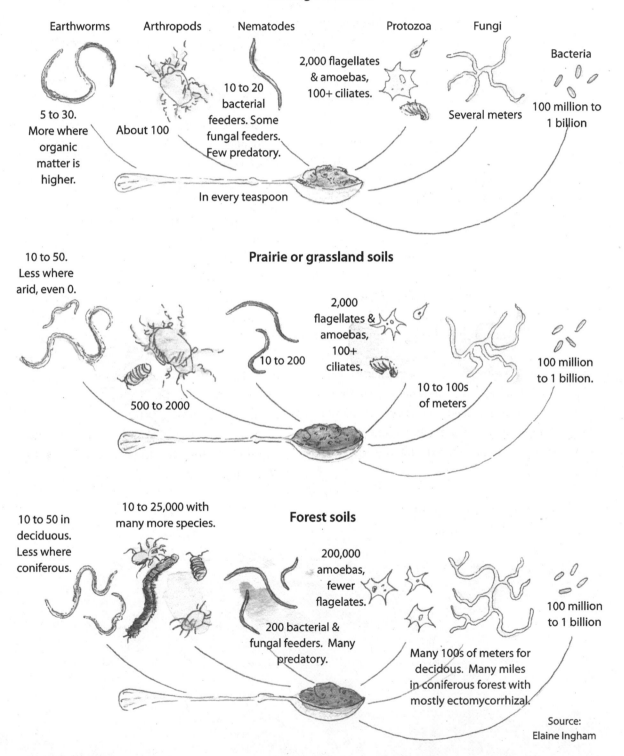

Average farm soil

Earthworms

5 to 30. More where organic matter is higher.

Arthropods

About 100

Nematodes

10 to 20 bacterial feeders. Some fungal feeders. Few predatory.

In every teaspoon

Protozoa

2,000 flagellates & amoebas, 100+ ciliates.

Fungi

Several meters

Bacteria

100 million to 1 billion

Prairie or grassland soils

10 to 50. Less where arid, even 0.

500 to 2000

10 to 200

2,000 flagellates & amoebas, 100+ ciliates.

10 to 100s of meters

100 million to 1 billion.

Forest soils

10 to 50 in deciduous. Less where coniferous.

10 to 25,000 with many more species.

200 bacterial & fungal feeders. Many predatory.

200,000 amoebas, fewer flagelates.

Many 100s of meters for decidous. Many miles in coniferous forest with mostly ectomycorrhizal.

100 million to 1 billion

Source: Elaine Ingham

CASE STUDY
Cucurbits and Mycorrhizal Fungi

There is no longer any doubt that mycorrhizal fungi have a large role to play in the efficient uptake of nutrients for crops. Many of our crops have evolved over immense spans of time in symbiosis with these fungi. However, it is also clear that not all crops form relationships with fungi (notably the brassicas), and these serve as a poor host for the next crop looking for quick access to fungal networks.

In addition, there is a diversity of fungal genetics at play, and some plant varieties and fungal strains companion better. Research with different melon cultivars and *arbuscular mycorrhizal* fungi showed that although

GREEN THUMB TECHNIQUE

Cover-cropped paths help maintain fungal populations adjacent to permabeds even if bed crop is not a host.

many combinations increased crop yield, some had greater effect.

Research of best mycorrhizal/crop symbiosis is an admirable horizon. For now, improving soil life and inoculating with a plethora of mycorrhizal fungi and helping them persist in the soil are well worth the investment. Crop rotation design should consider the seasonal continuity of mycorrhizal hosting for future crops. As such we focus on conserving hyphae (the branching filaments of fungi) through reduced tillage and inoculating after brassica crops as a rule of thumb.

Do You Know?

Soil life recolonization focuses on minimizing distance between healthy networks of soil life and newly prepared land for crops as well as a crop rotation that accounts for continuity in soil life habitat. As we will discuss, a *soil life conservation core* within each garden bed is fundamental.

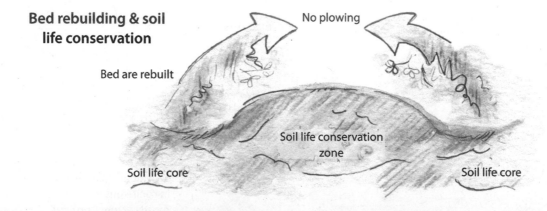

Bed rebuilding & soil life conservation

No plowing

Bed are rebuilt

Soil life conservation zone

Soil life core

Soil life core

Destroy Your Soil 101

Seeing how important soil is, what should we avoid? Destructive practices include:

Regular cycles of plowing, discing, and cultivating

- Inverts soil profile (protective layer is buried)
- Chops and compacts soil structure
- Disturbs soil life and reduces relationship connections

Excessive cultivation

- Oxygenates soil and burns up organic matter
- Destroys soil tilth

Excessive and deep tillage

- Destroys soil food web
- Chops up fungal hyphae
- Disturbs biopores
- Reduces connectivity of soil

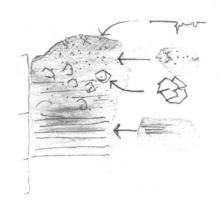

Compaction from machinery, humans

- Reduces pore space in soil
- Makes soil impermeable to movement of water and air

Removal of organic material

- Continuous removal of organic material out of field
- Better to leave as much as possible on surface to protect soil over winter
- Debris is a source of organic matter for soil improvement, increases cation exchange capacity, improves drainage and feeds soil life

Working saturated soils, especially clay, in early spring or late fall

- Compacts soil
- Creates rock-like blocky aggregates over plow pan
- Creates ruts for runoff
- Reduces soil absorption capacity

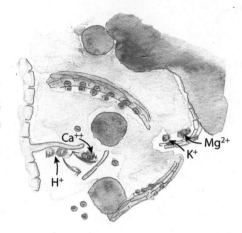

Cation exchange
capacity within a
soil aggregate

Ca^{++} Mg^{2+} K^+ H^+

- Crops become dependent on readily available nutrients and build less structure for deep nutrient mining, increasing vulnerability to nutrient loss.
- Excessive commercial nutrients can burn crops and soil life.

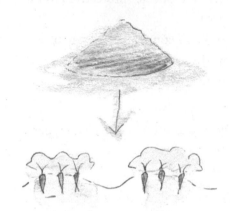

Regular periods of time when soil is bare/exposed to elements

- Rain compaction
- Erosion of soil
- Leaching and runoff of nutrients

Regular use of pesticides/herbicides

- Can kill soil life
- Promotes the evolution of super-pests and weeds
- Dirties the quality of water and air

Regular use of synthetic and on-demand fertilizers

- Makes us dependent on off-farm sources of fertility

- Find local and on-farm solutions for fertility

THIS ALL RESULTS IN:

Poor soil conditions

- Poor plant germination conditions
- Compaction layers where plant roots cannot grow deep
- Nutrients and water cannot move into soil
- Soil life cannot thrive and form connections with crops

Weak crops

- Delayed and slowed growth
- Restricted root growth
- Slow relationship forming with soil life
- Poor nutrient, water uptake
- Weakened cellular structure
- Weakened function of plant process
- Insufficient lipid formation in leaves

Crops vulnerable to

- Weather fluctuations and extremes
- Pest feeding and disease outbreaks
- Fragility due to human handling (opening wounds for disease)
- Weed competition
- Overmaturity but underdeveloped (old and small)
- Partial decay or death

Improve Your Soil 101

Simple Actions to Improve Soil

We will go deeper into soil improvement in our permabeds section. For now, consider these simple actions:

- Prioritize soil cover with crop or cover crop through seasons
- Winter/spring cover crops protect soil life against excessive cold and are established in spring to prevent erosion of soil and nutrients
- Summer/fall cover crops protect against excessive heat and moisture loss, and heavy rains
- Use green manures, compost and other sources of slow-release nutrients plus organic matter
- Inoculate soils with beneficial mycorrhizal fungi to form symbiotic relationships with your crops
- Inoculate with other soil micro-organisms to help invigorate the soil food web. Fresh manure in your rotation is a great source of biological activity for the pedosphere.
- Work the soil well in advance to avoid last-minute work in poor conditions
- Conservatively work soil (see S is for Soil, page192)

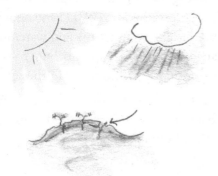

Ecosystem Services

Humankind has always benefited from surrounding ecosystems. *Ecosystem services* is the name given to all benefits we receive from the interaction of the living and non-living entities and forces of Earth. Their nutrient cycles and energy flows sustain us in many ways.

It is easy for us to see some ecosystem services. A forest for its production of lumber, a mountain for its minerals and rivers for hydro-electricity. Looking deeper there are so many more services revealed. We must also learn to farm for the enhanced services it provides.

Four Types of Services

- Provisioning: Relating to the production of resources (food, water)

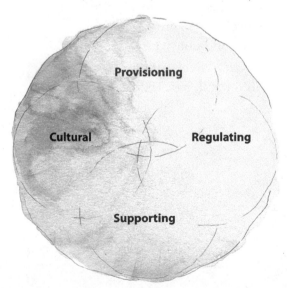

- Regulating: Relating to the control of disease, weather
- Supporting: Good examples include nutrient cycling, or crop pollination
- Cultural: We have fun in the sun and peace in the woods (recreational and spiritual benefits)

Farm Ecosystem Services

Let's look at some of the services farmers receive.

Energy flows are all around us.

Provisioning: Soil life interacts with heat and moisture to provide the service of decomposition and provisioning of soil organic matter. Many plants, animals, fungi, etc. provide us with food.

Regulating: Trees and other plants provide microclimates, sheltering against extreme rain, sun and flood.

Supporting: Mycorrhizal fungi form symbiotic relationships with plants and allow increased access to nutrients and water in exchange for sugar. They contribute to the cycling of nutrients and water in the soil.

Cultural: Healthy diverse food systems provide habitat, beautiful spaces and research/educational opportunities.

Assigning Economic Values

We need to start to understand the value of ecosystem services. Let us consider the questions:

- What is the value of biologically active soil?
- What is the value of topsoil retention against erosion?
- What is the value of habitats that sustain healthy predatory insect populations?
- What is the value of a beautiful and functional landscape that inspires farmers?
- What is the value of a tree that regulates moisture, holds and builds soil and provides nuts, lumber and mulch?

Trees provide many services: release oxygen, sequester carbon, build soil organic matter, stabilize soil, provide habitat for beneficial organisms, etc. They are the biggest organism on our planet, connected together through mycorrhizal fungi.

Just geocaches!

Do you take credit?

Pando is the name given to a 104-acre clonal colony of a single male aspen in Utah whose root system is estimated at 80,000 years old.

Seasonal Rhythms

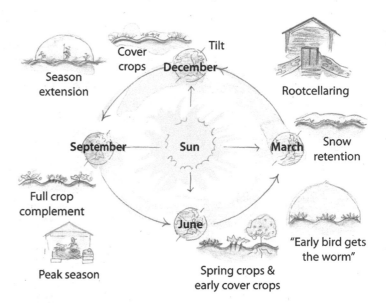

Cover crops

December

Tilt

Season extension

Rootcellaring

September

Sun

March

Snow retention

Full crop complement

June

"Early bird gets the worm"

Peak season

Spring crops & early cover crops

Do You Know?

We need to value these functioning systems and support them. We wouldn't expect our mechanic to repair the car for free. We wouldn't expect the stove to run without gas. We wouldn't assume to walk without feet.

The ecosystem services that are already essential to farming can provide us with so much more once we give them space to function and support them through *best production practices*.

Ecosystem Services

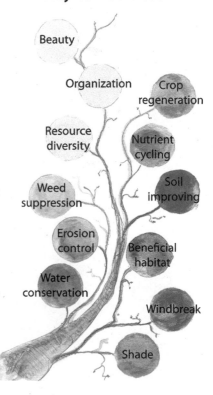

Beauty

Organization

Crop regeneration

Resource diversity

Nutrient cycling

Weed suppression

Soil improving

Erosion control

Beneficial habitat

Water conservation

Windbreak

Shade

Support Ecosystem Services on the Farm

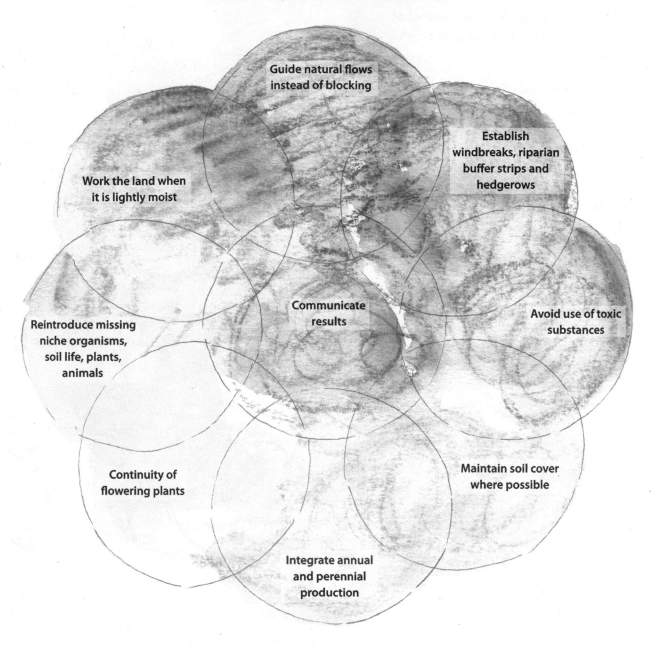

Guide natural flows instead of blocking

Establish windbreaks, riparian buffer strips and hedgerows

Work the land when it is lightly moist

Reintroduce missing niche organisms, soil life, plants, animals

Communicate results

Avoid use of toxic substances

Continuity of flowering plants

Maintain soil cover where possible

Integrate annual and perennial production

CASE STUDY

Ecosystem Services: Hover Fly Larva Predation of Aphids in Lettuce Production

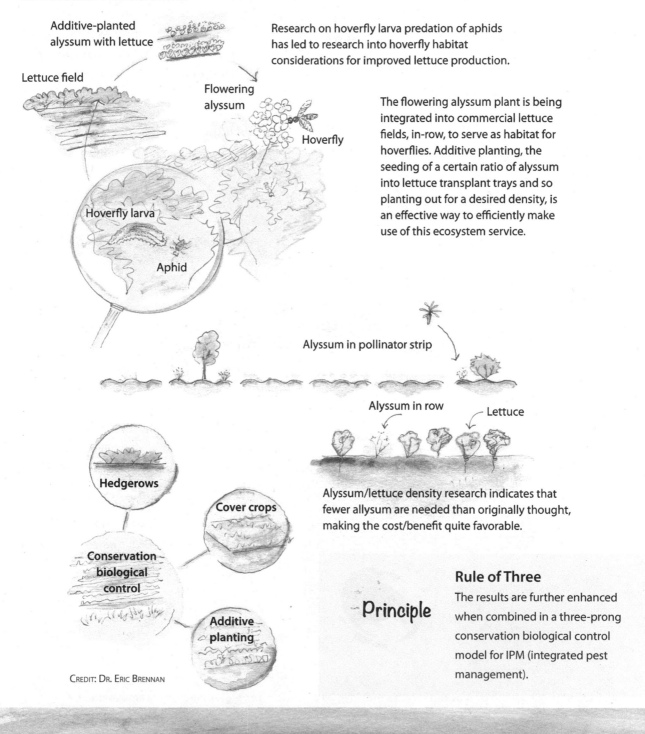

Additive-planted alyssum with lettuce

Lettuce field

Flowering alyssum

Hoverfly

Hoverfly larva

Aphid

Research on hoverfly larva predation of aphids has led to research into hoverfly habitat considerations for improved lettuce production.

The flowering alyssum plant is being integrated into commercial lettuce fields, in-row, to serve as habitat for hoverflies. Additive planting, the seeding of a certain ratio of alyssum into lettuce transplant trays and so planting out for a desired density, is an effective way to efficiently make use of this ecosystem service.

Alyssum in pollinator strip

Alyssum in row

Lettuce

Alyssum/lettuce density research indicates that fewer allysum are needed than originally thought, making the cost/benefit quite favorable.

Hedgerows

Cover crops

Conservation biological control

Additive planting

CREDIT: DR. ERIC BRENNAN

Principle

Rule of Three

The results are further enhanced when combined in a three-prong conservation biological control model for IPM (integrated pest management).

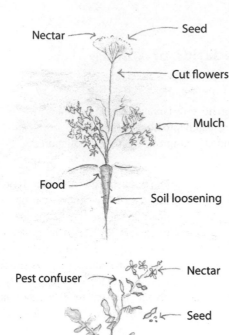

Nectar — Seed

Cut flowers

Mulch

Food

Soil loosening

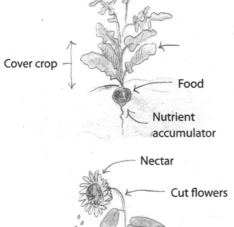

Pest confuser — Nectar

Seed

Cover crop

Food

Nutrient accumulator

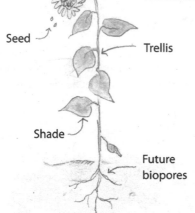

Nectar

Cut flowers

Seed

Trellis

Shade

Future biopores

Crop Life Cycle and Services

Crop Services

Taking any crop we grow — carrot, radish or sunflower — we begin to observe its services. We often think of crops as having one service: food, when in fact they have many. To truly comprehend the service potential, consider its morphology, habit, niche and life cycle. These give clues to how it fits into our agricultural designs. Now that we have looked at crop life cycles, let's look closer at crop services.

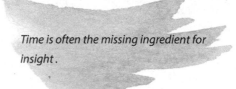

Time is often the missing ingredient for insight.

Ask Some Questions!

Crops are more than just food, they are plants in their own right with their own character, and plant character reveals service potential.

Ask these questions. Is this plant an annual, biannual, perennial? What are the characteristics of its roots, stem, leaves, flowers, fruits? How would it disperse its seeds in nature? What plants might we find in association with this plant in a natural system? How much organic matter does it produce? What weeds does it out-compete? What other products can it produce?

Hidden yields are the overlooked services provided by a crop, such as organic matter accumulation or windbreaking.

Life Cycle Consideration

A crop is not a stagnant entity, existing only as the mature food-producing unit. Farmers will be familiar with many of the stages of growth up to the production of the edible portion of the plant. Yet, it is the full understanding of a crop's *complete life cycle*, when left on its own, that is of interest here. This cycle allows us a view of potential crop services that may be overlooked when considering only plant germination, growth and production of the food part.

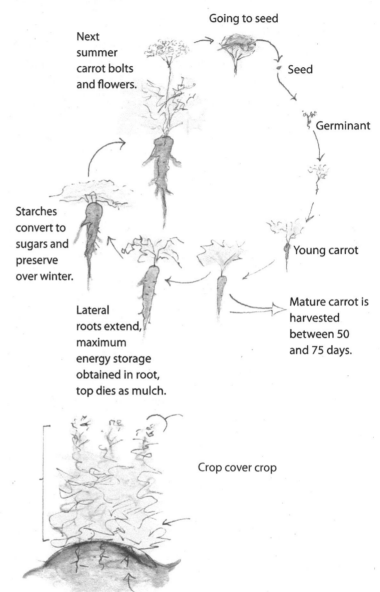

Going to seed

Next summer carrot bolts and flowers.

Seed

Germinant

Starches convert to sugars and preserve over winter.

Young carrot

Lateral roots extend, maximum energy storage obtained in root, top dies as mulch.

Mature carrot is harvested between 50 and 75 days.

"Hey, I am more than just food, you know!"

Crop cover crop

It is the missing stages that are needed for a full comprehension of a plant's *services* and *hidden yields*. For instance, only by observing a plant going to seed did humans realize they could collect the seed and bring it near to their homes for cultivation. Furthermore, even though we don't collect lettuce seed on our farm, the process of letting it bolt and flower has become an invaluable tool. We use it as a *crop-cover crop*, *habitat hotel* and *in situ* mulch production. Observing the organic matter production, habitat buzzing and weed eradication potential of an overgrown bed of greens gave us perspective into the easy service of giving the plant a little more time to grow.

Crop-cover crop means a crop that is left to build organic matter, fix nutrients and provide cover.

Agro-ecosystems

Because ecosystems are efficient at converting and moving energy, mimicking them is key for market gardens. Consider the fact that your garden is already an ecosystem and look for ways to make it function more like one. What are the inputs to our food systems and what are the outputs? Do the different species fit together? How do the species serve each other and the greater production? Our

Ecosystem Dynamics

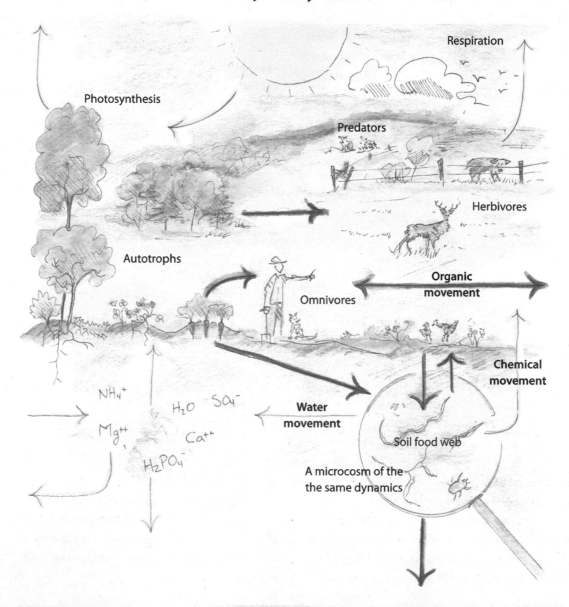

Agro-ecosystem, agro-ecology, ecosystem farming. In other words, growing like an ecosystem: a layered and diverse agriculture allowed to evolve and regenerate for multiple services, including food, and maybe fodder, fiber, fuel, or even soil function-support, community beautification and water purification.

biggest question being what ecosystem do we want to mimic? We decide we want a woodland market garden; now, with an intended agro-ecosystem in mind, we can begin to design.

Managing an Ecosystem

An ecosystem is complex, yet, for the market gardener, this must be more organized and more controlled. There has to be profit, and we don't want to farm a permaculture market garden mess. We should find a balance of diversity, maximize crop services and manage specific situations in a standardized way. We talk about this in our *permabed system* and the way we use permanent raised beds to help manage diversity. We also will look at *organizational land patterning* and how we can find various solutions to dividing up our land for better management of diversity throughout the farm.

It is critical, however, that we choose the initial species that make up our ecosystems. Species with benefits we are willing to manifest. For example, in a savannah market garden, grass would be an asset as it is eaten by the herds of animals that are rotated through. On the other hand, if your primary production is vegetables and mixed fruits and berries, many grasses are less beneficial and more an invasive weed. In an agro-ecosystem where regeneration is encouraged, choose species that you won't mind going wild.

Agro-ecosystems, ecosystem farming, agro-ecological production…growing like an ecosystem!

Support Ecosystem Services

We operate as though Earth's bounty is endless. Yet, the soil can be lost, and we cannot recreate the wealth of nature in our lifetimes. Soil is formed on a geological time frame. We must understand the ecosystems that perpetually create wealth and harness their services to maintain our soils, waters and biology so they can serve our needs.

However, we cannot reject our current agricultural models completely. We must evolve, as nature does, in the direction of connectivity. Understanding and valuing ecosystem services is not enough. Now let's design them into our market gardens and fuel them as we would a tractor. Sow cover crops for winter so soil life can sustain spring production. Support the systems that would support us. Work the land carefully because it is precious.

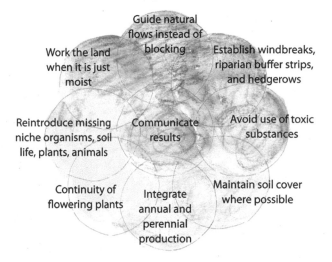

Guide natural flows instead of blocking

Work the land when it is just moist

Establish windbreaks, riparian buffer strips, and hedgerows

Reintroduce missing niche organisms, soil life, plants, animals

Communicate results

Avoid use of toxic substances

Continuity of flowering plants

Integrate annual and perennial production

Maintain soil cover where possible

Principle

Organizational land patterning is the concept of dividing space into well-defined places. Defining land into fields, hedges, plots and raised beds can help integrate the diversity needed for ecosystem farming. Once space becomes place, we can pattern it for better management of diverse production.

Support Farm Ecological Succession and Agro-ecological Evolution

More than anything, permaculture farmers should work to support farm ecological succession. Instead of constantly keeping our farm in a perpetual state of disturbance, brought by constant tillage of all parcels of land, we should set aside land through organizational land patterning to continue down a more natural ecological trajectory toward woodlands. For instance, designating a regular pattern of beds in the garden for perennials is a powerful tool.

Key Ways to Support Garden Ecosystem Services

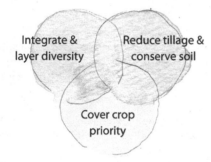

Integrate & layer diversity

Reduce tillage & conserve soil

Cover crop priority

1. Integrate Diversity

- Integrate annuals and perennials, cover crops
- Integrate crops and cover crops
- Integrate crops into guilds
  - Creates garden microclimates, shade/shelter/protection
  - Improves nutrient cycling with healthier soil
  - Improves intercrop services like pest deterence

Life Cycle of Parasitic Wasps

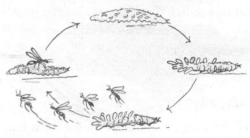

Out of the long list of nature's gifts to man, none is perhaps so utterly essential to human life as soil.

— Hugh Hamond Bennet

2. Reduce Tillage

- Put garden into permanent raised beds
- Practice conservation tillage techniques
- Till shallow, seldom, softly and stratified
  - Reduces loss of carbon from excess soil aeration
  - Protects mycorrhizal and other soil life
  - Reduces plow pan and improves soil structure

3. Cover Crop Priority

- Allow crops to become cover crops
- Cover crop spent beds soon and often
- Ensure cover crop over winter and spring
  - Builds organic matter and nutrients
  - Protects crops, soil life and soil tilth
  - Reduces erosion and nutrient loss

These three simple ways of supporting ecosystem services will pay you back quickly. We have seen much higher yields by reducing tillage, creating shelter and increasing beneficial insect habitat.

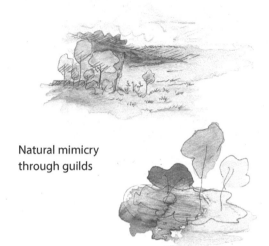

Natural mimicry through guilds

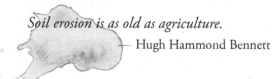

Primary services include shade/shelter/ protection, especially of soil tilth, growing crops and soil life. We fight to keep the top 2" of our soil wet in summer to germinate carrots. In a woodland market garden, the evaporation would be less, the organic matter higher and the water needs for germination better met. We could achieve better stands of fall carrots.

Unnatural Operations that Counter Ecosystem Services

- Constant soil disturbance
  - Although disturbances occur in ecosystems, they don't occur every single year on as grand a scale as agriculture employs. often, this causes *succession stagnation* and prevents any building of complexity needed for *farm ecosystem functions.*

- Focused eradication of select pest species
  - Regular and full eradication of species is ecologically abnormal and leads to the very scary outcome of super bugs!

- Incomplete life cycles of species and imported genetics of dominant species
  - Such large tracts of land populated by communities that have no local adapted genetics are abnormal. Working to slowly reintroduce seed saving into every farmscape at multiple scales is critical.

Soil erosion is as old as agriculture.
— Hugh Hammond Bennett

GREEN THUMB TECHNIQUE

Many predatory insects feed on pollen and nectar in their adult stage; provide this by designing continuous bloom into your garden.

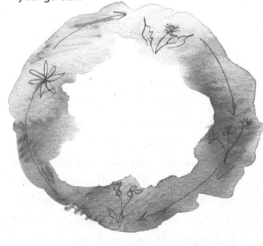

Succession stagnation occurs when disturbance, such as plowing occurs annually. As you can see, most agricultural fields are perpetually kept in the earliest stages of natural succession: bare exposed earth and annuals.

EXAMPLE: Provide for the Needs of Beneficial Insects

Beneficial insects require food, shelter and water like other creatures. Through cover cropping, shelter belts, hedgerows and polyculture, we provide a reliable source of pollen, nectar and protection. However, many *beneficials* are predators, requiring quantities of herbivorous insects as food. I guess that is what makes them beneficial — they eat those guys that eat our plants.

Let's switch perspective and regard pests as a reliable food for the *beneficials*. It is when we achieve economic yields of crops without upsetting this natural prey/predator relationship that we build a stronger insect defense on the farm. Use of pesticides upsets this balance by both accidentally killing beneficials and removing the food source they need for population cycles.

Doctor said: "50 aphids a day will keep the pests away."

That being said we shouldn't let our crops be destroyed. Using row cover, trap crops, delayed planting and other IPM techniques helps keep crops healthy while engaging in *insect husbandry* (that purposeful feeding of pests with trap crops to raise them as food for their predators). We let a bed of kale over winter on the complete opposite side of the field from where we plan to plant early brassicas. This is very effective at trapping flea beetles!

In Conclusion

Many ecosystem services will not be immediately apparent but will reveal themselves through key strategies that enhance garden ecology. By reducing tillage, integrating diversity, and prioritizing cover crops, we build ecological integrity. Through observing crop life cycles and gaining a better understanding of a crop's character, we gain insight into their hidden yields and can begin to design them for increased service. A solid basis of natural sciences can help us in our inquiry and design of the ecosystem farm.

GREEN THUMB TECHNIQUE

Old seed with unreliable germination can be used for a *pest cocktail* to build habitat hotels. This mix can be broadcast on a bed to attract a range of organisms. The mix is conveniently an assortment of most vegetables you grow and so will attract a range of pests. As this is seeded early and left exposed, it is an ideal feeding frenzy. The pests are happy to stay here and be eaten. It is tempting to go spray the heck out of them as they are concentrated, but I would discourage this as it interrupts the beneficial organism husbandry and will result in collateral damage.

Habitat hotel

ETHICS

Once we understand and accept farms as within nature and begin to put value and even ($) on *all* of Earth's wealth, direct and indirect, we can begin designing the *ecosystem farm*, managing it with *design management* so it may evolve as it is only right for a living production. Understanding that we need to support the ecosystem so it can serve us, and maintaining it like any other tool.

However, we must not only recognize, understand and value ecosystem services, but also show appreciation. This closes the loop between profiteering and *profit resilience*. The permaculture ethic — people care, earth care and surplus share — are an inspiration to us. They ask us to remember our place within our communities and ecology and share with each other to grow forward.

Thanksgiving to Ecosystem Services

As a guiding way of regularly demonstrating appreciation, we use the thanksgiving address.

The Iroquois Nation, a confederation of Native American peoples, whose sociopolitical model contributed to modern North American democracy, has a ritual of giving thanks to all life.

I have found that regular thanksgiving enhances my remembrance of the benefits of Earth's systems and connects me to the origins of my design management.

Our moon, sun, and Earth were here long before us, so it only makes sense that they be given the respect and familial titles that are traditional. However, we need not feel too hippy-dippy about this; it is a practical presentation of those entities that put money in the farmer's pocket and make our world enjoyable to live in.

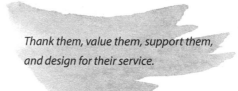

Thank them, value them, support them, and design for their service.

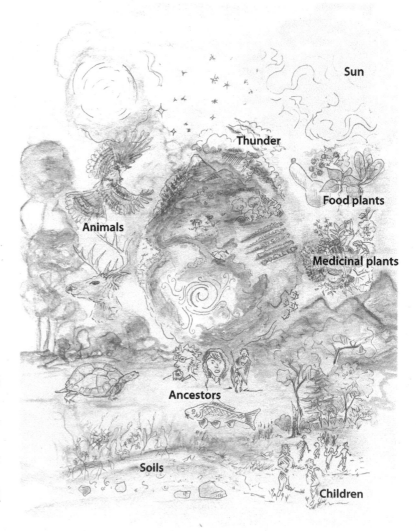

THANKSGIVING ADDRESS

We now give thanks to:

Grandmother Moon
Grandfather Sun
Father Sky
Mother Earth
The stars
The four winds
The thunder clouds
The food plants
The medicine plants
The wild plants
The trees
The rocks
The soils
The waters
The fish
The animals
The birds
The insects
The elders
The children

If there is anything that I have not set in words, I hope the illustrations capture the grandeur; if there is anything unseen, please give thanks in your own way.

Now our minds are one.

*Inspired by the Haudenosaunee Thanksgiving Address

Whole Farm Mapping

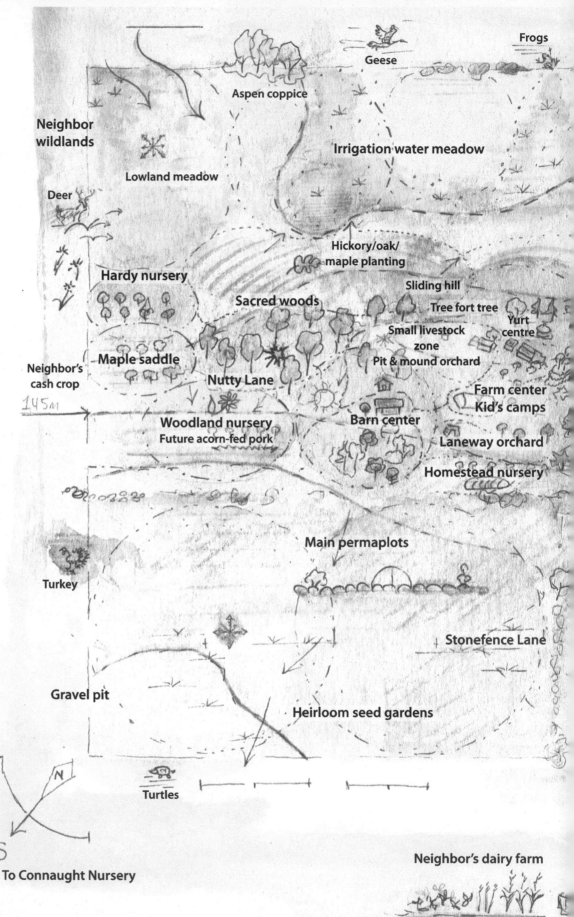

Whole Farm Mapping for Design Management

Frogs

Geese

Aspen coppice

Neighbor wildlands

Irrigation water meadow

Lowland meadow

Deer

Hickory/oak/ maple planting

Hardy nursery

Sliding hill

Sacred woods

Tree fort tree

Yurt centre

Small livestock zone

Maple saddle

Pit & mound orchard

Neighbor's cash crop

145m

Nutty Lane

Barn center

Farm center

Kid's camps

Woodland nursery
Future acorn-fed pork

Laneway orchard

Homestead nursery

Main permaplots

Turkey

Stonefence Lane

Gravel pit

Heirloom seed gardens

N

S

Turtles

To Connaught Nursery

Neighbor's dairy farm

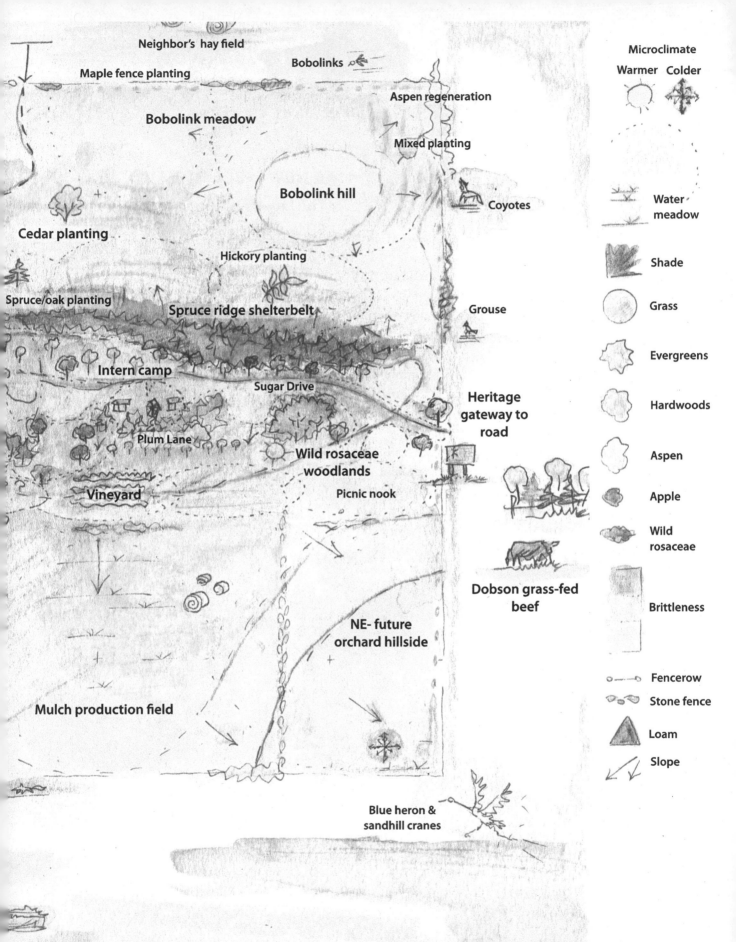

Why Map Your Farm?

We look closely at the site itself, thinking of everything as a potential resource (clay, rock, weeds, animals, insects).

— Bill Mollison (*A Designer's Manual*)

I have been mapping since I was a little kid, starting with treasure maps of my backyard. I mapped to find my treasure! Arrrr! Yet, your treasure be all around ye!— In the soil, in the currents of air, in the interaction with your farm ecosystem and community. It just requires opening your mind, printing a property map and making some notes through the seasons.

By mapping your property, you can begin to organize your farm for efficient production. Mapping helps you lay out your land according to what is already there. **Understand the lay of the land before you lay out anything on it.**

A NEW PIECE OF LAND

The Layers of the Land

Before mapping your land, you must understand your bioregion: Boreal, mixed woodland, temperate forest, Carolinian, etc. This gives you a broad view of your ecological context.

We should also understand the fundamental layers of any piece of land.

Your land has layers! We can isolate the indivisible aspects of your piece of land into distinct categories to help us better see how they interact. There are 12 layers that will help you observe, map and plan your property.

Mapping Your Farm Environment
(Layers of the Land)

Climate
- heat regime
- moisture regime
- weather & prevailing winds
- hardiness zone
- microclimates

Macrofauna
- wildlife
- domestic animals
- birds, mammals, amphibians, reptiles

Geology
- bedrock
- outcrops
- unconsolidated materials
- parent materials for soils

Micro Fauna
Insects, Fungi & Micro-ecology
- pollinators
- top predators
- pests
- beneficials
- mycorrhizae
- other edible fungi
- mildews
- bacteria
- protozoa

Topography/Geomorphology
- elevations
- highest and lowest points
- slope degree
- complex or simple
- consider aspect
- gravity irrigation potential

Society
- common purpose
- community interactions
- ethnicity
- distance
- product supply and demand
- local economy

Hydrology
- precipitation
- streams, ponds
- water table

Brittleness
- seasonal wet spots
- marshes, water meadows, etc.
- soil water-holding capacity

Infrastructure
- roads
- outbuildings
- domiciles
- utilities

Soil
- parent materials
- profile & horizons
- structure
- texture
- stoniness
- organic matter
- soil life
- overall integrity

Flows
- human
- animal
- wind
- water

Vegetation Ecology
- spruce ridge
- water meadow
- oak maple bush

Ecosystem Dynamics
- interaction of all layers

Finding Potential Profit Centers

Profit centers are micro-enterprises within your farm business. Farms should focus their production on a few key enterprises that are suited to the farm's environmental layers and each other. It doesn't make sense to plow up steep, stoney ridges for gardens or plant orchards in low, wet meadows! It also doesn't make sense to choose enterprises that are unable to help each other. We will talk about this later, but for now, consider how farm enterprises suitable to your land will also benefit each other. Keep this in mind when walking the land and looking for production opportunities. We are looking for three harmonious productions; harmonious to the land and each other.

Walk your land and look for clues for suitable production on *this* land. Consider all 12 layers when observing a property, but focus on these three key areas for design:

1. Resource ecologies
2. Microclimates
3. Flow dynamics

Look for key features. You might see an opportunity for a market garden, a cattle herd or a compost production business.

You may find a woodlot suitable for one type of production and a field for another. While assessing these, keep the following in mind:

1. Your goals, skills and resources
2. Your community resources and potential clientele
3. What your future would look like running these potential enterprises

What is needed to see these opportunities into a steady-state business?

We will discuss this further in the next section on holistic planning. But before you can truly plan, you must have an idea of the land you plan to farm and see what it has to offer.

Ecological Land Classification

Your Local Ecologies

Ecological land classification is a system developed to help classify and name distinct ecologies that reoccur with frequency. Consider reading up on your local ecologies to better understand the farm ecosystem. You can find information in textbooks, local databases, research papers and periodicals. You can also talk with neighbors, naturalists and local professors to better understand your ecologies.

Principle

Rule of Three

We consider three as a unifying design principle because design should have both a motivation for more diversity and a limit to it. Three works well because it is the first number where this complex interaction and true community exist.

Do You Know?

Brittleness is a measure of the ecosystem's moisture dynamics, with one extreme being desert and the other tropical rainforest. In our area, ecological brittleness looks like this:

bog → water meadow → meadow → upland field → gravelly field → gravel pit

---

The ecology of your property is foremost in decision-making and profit center selection. Look over your holistic goal and use the decision-making test questions to help you. Make sure the roads you open physically and metaphorically lead you where you have planned to go!

The Cultural Landscape

It is also important to make note of the cultural landscape when assessing a property: the landscape humans have built — barns, roads, wells, fences, etc. Make note of features that are useful, needing repair and possible obstacles to your plans.

Key Features: Property (design) features include specific environmental and cultural items: a barn, a fenced paddock or a tall oak in the middle of a field. Features are important for mapping because they often act as foci for sketching, observing and patterning the land. For design, they act as key features to design around because they are either immovable, valuable or constitute a preexisting or dominant pattern to follow. Human-made features are often key as they have much imbued value and delineate spaces. Key features can be considered dominant features, like a central ridgeline or complex of barns.

What is your ecology?

Geology: acidic clay basin, alkali glacial moraine, residual brown clay loam
Hydrology: wet water meadows/dry hillsides, well-drained fields with pockets of seepage
Plant Ecology: acidic/wet grasslands, upland mixed deciduous forest, nutrient-deficient hay fields
Animal Ecology: cattle, deer, coyotes, geese, humans, grouse, chipmunks, pigs, chickens, insects, voles

Geology: flat bedrock substrate with shallow soil, exposed bedrock islands, increasingly deeper soil
Hydrology: poor drainage from impermeable layer, overland flow off-island, better draining in deep soil

Geology: rich mucky soils, drainage ditch
Hydrology: poorly draining, better with ditch, best near forested edge where trees have opened fissures in the underlying clay

Geology: fine sandy land above folded undulating rock
Hydrology: creek running to lake
Plant Ecology: cattails, drought-resistant grasses, upland mixed pine forests
Animal Ecology: wood ducks, otters, wolves, bears, goats, humans

Familiarize yourself with YOUR landscape

Extended crop services are services provided by a crop after its primary purposes have been met. These are often achieved by leaving the crop to become a crop-cover crop and include soil protection, organic matter production and habitat.

OBSERVATION

When approaching whole farm mapping, it is important to take stock of all the resources at your disposal. These resources hidden within the layers of your land, help form your resource base. Your base also includes your personal, family and community resources. But for now, let's look at how we can observe the land for its useful entities.

Map Your Property's Resource Base

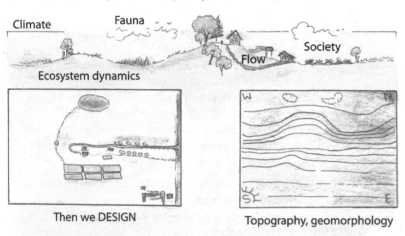

Climate · Fauna · Society · Flow
Ecosystem dynamics

Then we DESIGN

Topography, geomorphology

Vegetation ecologies, brittleness

Hydrology

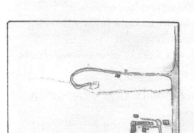

Infrastructure

Pedology & endopedology

You are looking for potential areas to design ecological production or place needed infrastructure and projects. Consider which areas are good for gardens, pastures, ponds and barns, houses and driveways.

Observational Mapping

Observational mapping requires walking a property and noting key features, ecologies, and flows and recording them in map format, showing the lay of the land. This may include a NE loam hillside, an old barn with good wooden structure but a bad foundation and prevailing winds. Taking stock of these resources on a map shows their value: utility and obstacle or benefit in relation to each other across a property. Often, it helps to use a base map, to record observations made from a property map. However, it can be as simple as walking and taking notes and flagging sites to begin to familiarize yourself with the land: its water, its soil, its winds, its snows and even the way animals move across it.

Base Maps

Base maps are needed to help this process roll along. They can be made from property maps but can also include information from other map resources found online or through municipal, regional and global databases. You can start with a simple property map showing roads, buildings, utilities, property lines. A map with layers showing vegetation, water and elevations is also useful. Adding a map showing local soils is very helpful. Photocopy these and use them for making observations and drafting designs. You can integrate important features, ecologies and flow on one map or have them layered separately. Base maps will be used to draft infrastructure, production areas and other projects. Eventually they can form your permaculture property map. You can also do this using graph paper, Excel spreadsheets, Google SketchUp, GIS and other mapping software.

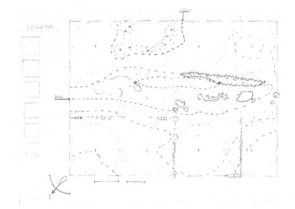

What You Can Find with Observational Mapping

1. Microclimates to shelter animals, harbor hardy perennials and nurture niche crops
2. Rich soil pockets for your initial garden, good soils to nurture into future gardens
3. Water flows to gather into reservoirs, move for irrigation and nurture for habitat
4. Special plants to propagate in nurseries, plant for windbreaks and grow as crops
5. Fauna that may be a resource or require special crop protection
6. Insect/disease/weed populations to balance

7. Fungi to propagate for food, inoculate for soil health
8. Your place on the land, placement of structures, the flow of people onto your land
9. The current infrastructure's potential as it pertains to the plan you are setting in action. Where you will grow, process, store, cure, etc.

Maps need not be complex; simple notes on a photocopied property map is often best. The act of regularly walking and spatially recording observations is what is most powerful.

Walking the Land

Walking the land is the process of moving through an entire property. Do it slowly, methodically and take notes. I take notes on base maps of the property, in notebooks and make voice memos. Walk the fence lines and see how ecologies continue past them into your neighbor's land. Follow contours, animal paths and other natural ways to see how the land flows. Discover key places and think about their potential for your farm business. In order to do a good job, you must observe spherically.

GREEN THUMB TECHNIQUE
_____

Make a legend so you can index key features, ecologies and flows in the margins.

Garden Mapping: Observations, Planning and Records

There are certainly many ways to map, and here is one easy model for simple observations, planning and records using Excel, Numbers or similar software. A base map (like the one provided) shows three main fields (A, B, C), each with six garden plots. These plots, when delineated this way, are useful for recording all sorts of data.

We use these:

- Environment mapping, noting vegetation, moisture regimes, soil texture, microclimates, etc.
- Seasonal observations, indicating weed pressures, surface erosion and water flows, snow drifting, etc.
- Chart-planned production rotations, marking crops, planting dates and bed feet into the Excel squares. This is useful for organic certifiers and for your own crop planning needs.

Making an Excel base map:

- Assign a value to a graph square. We usually make it either our chosen organizational unit (6'×300'), or (6'×25' or 6'×50'), so that each square is either the equivalent of one of our permanent raised beds or a fraction of one bed.
- Organize your garden layout using these organizational units (see page 86).
- Make sure to organize for lanes, roads and even infrastructure.
- It is not important that these be exactly to scale, only that they serve as a framework for you to make notes and records.
- You now have a base map, which means you can print many copies for field observational walkabouts (see page 158), and can add pertinent data to your digital copies when needed.

C Environment Mapping — WET! / Frost pocket / Loam / Stony

B Seasonal Observations — ENCROACHING GRASSES / ← Compaction → / low fertility

A Planned Production — Squash / Low weeds / Carrots / Seed cover crop / Rye & clover / Mixed veg / Turn over! / Plant hedge next year.

Using color:

- Black and white maps and colored pencils can combine nicely for seasonal observations. As you make observations, shade in areas with green to show extent of perennial grass encroachment, blue to show seasonal wet spots or erosion channels and red for areas of excessive stoniness.

- The mix of rigid Excel graph square and free shaded colored pencil can really aid the process of organizing records of these very fluid natural phenomena.

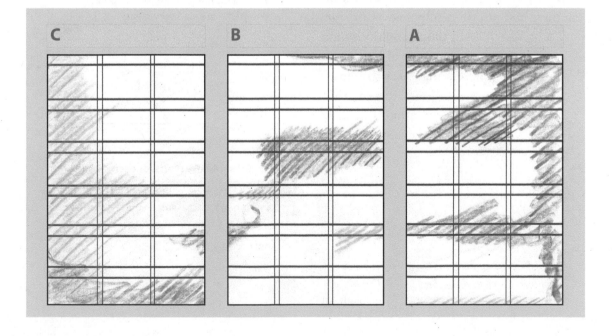

C

	Heavy clay	
Wet		
Frost pocket		
Loam		
Stony		

B

Encroaching grasses		
	Compaction	
Infertile area		

A

Curcubits		
Carrots		
Rye & Red Clover		
Early Mixed Veg		
Perenial hedge		

C   B   A

Spherical Observation

While walking your land, remember to open the mind's eye. My dad says we can see spherically. I agree! *Spherical observation* means we aren't limiting perception to the eyes. Let your eyes focus and marvel on the details; then back out for a global picture and property context. Then, re-focus and take in the minutiae with a new perspective.

Also, open your ears and listen to the wind and hum of insects, then make assessments. Feel with your hands; the soil is wet and crumbly here in JULY. We also watch the way greater patterns play out across the land. The way the wind blows over the fence from our northern neighbor's open fields. The way the deer trails lead back to that bush lot over to the west. The way the color of the grass changes, yellowing up a gravelly hill and greening down into the water meadow.

Spherical observation is key to seeing the land as it is, rather than as your mind may make it. This means you can see its potential services and obstacles, rather than disregarding these and just drop-shipping your dreams and aspirations onto a piece of Earth. See the places that are already there in this space and then make them your own.

Observe nature thoroughly rather than labour thoughtlessly.

— Bill Mollison

Seasonal Observation

Since we walkabout in every season, it is a good opportunity to make seasonal comparisons and to map and observe phenomena only available in that given season. This often leads to critical data for decisionmaking and farm design.

Early Spring: Best time to see how water moves on the land, especially noting where it moves *off* the land. Make notes in your design book to consider slowing, capturing and storing it.

Summer: Best time to see the productivity of the land. Where the grass is greenest and tallest, where trees of the same species put on more growth, etc.

Fall: Best time to record frost pockets. Where do tender crops die earliest? Light, frosty autumn mornings are beautiful...so get out of bed and chart frost lines.

Winter: Best time to observe... snow drifting! Also record shading in design zones for potential buildings or gardens. Animal movements are also very clear.

GREEN THUMB TECHNIQUE

Use different base maps for each season and overlay last year's to see ongoing seasonal phenomena. This assures you — yes, that is where the snow drifts deepest.

MAPPING FOR DESIGN MANAGEMENT

Now we should turn our mapping exercise from observational mapping to one of active design management. We must map for management. Map it now, so it is easier to manage later. We must map for design. Map what we will have to work with for designing. First, understand the layers of the land, then organizational land patterning, and finally designate design management zones.

Key Farm Design

Features	Key Resource Ecologies	Flows	Microclimates
• Roads	• Woodlands	• Wind	• Cool/moist
• Utilities	• Meadowlands	• Rain	• Warm/dry
• Infrastructure	• Water meadows	• Snow	• Hot/humid
• Wells	• Fertile fields	• Light	• Dark/cool
• Organized production areas	• NE hillsides	• Heat/cold	• Sunny/warm
• Fencing	• Rock outcrops	• Human ways	• Winter shelter
		• Animal ways	• Summer shade

Using Maps for Design

PERMACULTURE PROPERTY MAPS

A full property map showing your productions, design zones and the significant natural phenomena they interact with.

PROPERTY MAPS

Any maps that familiarize your property, its ecologies and human landscape.

DESIGN MANAGEMENT ZONE

A map showing exactly the layout of a design management zone, including placement of structures, roads, etc., as well as the clear flow throughout the space and organized management strategies.

ORGANIZED BASE MAPS

A basic map of property limits, macroecologies, water resources, elevations and human landscape. Used for many other map layers. Map is delineated with organizational units.

PROJECT PROPOSAL MAPS

A map of a design area with proposed layout of projects. Could include buildings, roads, composts, fencelines, tree plantings.

SEASONAL OBSERVATION MAPS

A base map used for taking notes throughout the seasons. Includes shading in areas of observed flow (wind, water, snow, animals) and key points of interest for phenomena (mushroom, wild leeks or deep soil pockets and springs).

Organizational Land Patterning

OLP can take many recognizable forms: how we divide grassland into paddocks or cropland into fields. It also considers the layout of roads, hedges, lanes and paths and the inclusion of ponds, streams and reservoirs. It affects where we grow, build and dig. Land patterning is the pattern we leave across our terrain over seasons and generations.

Organizational land patterning is patterned layout for farm management.

Dividing Space
Organizational Land Patterning

Start with a few clear features like:

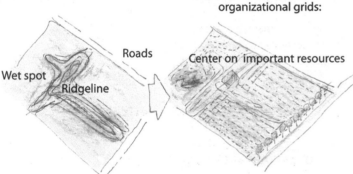

Wet spot
Roads
Ridgeline

Design around features, radiating organizational grids:

Center on important resources

Develop along most obvious lines

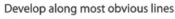

Add diverse production in management-ready places

See the land as organized space to make place

- Imagine your land overlaid by a uniform grid of organized space.
- Do a simple drawing of your farm on graph paper or take a property map and draw a grid overtop.
- Each cell of your land is a unique placeholder with resources needing management and organization. Defining these spaces into places makes management easier.

Choose an organizing unit for codifying your farm space

- Sometimes it is best to start by organizing the spaces of primary production — for us, the market garden — and then use this unit as a standard throughout the farm.
- Our organizing unit is 6'×300' because that is our standard raised bed. This standard was chosen because it fits well into our property dimensions and because many supplies, like row cover, are easily purchased in multiples of 300 feet.
- We organize garden beds at 6' wide, farm lanes at 18' and drives at 24', etc.
- Organizational units should consider the land, the tools, supplies and scale of production.

See your land as equivalents of your standard unit

- A beef farmer may see it as acres of pasture; we see it as 6'×300' sections.

Turn space into place using standard units

- Use standard units to lay out gardens, infrastructure or modify existing situations to your future needs.

Build out from key features, obvious lines and important resources using standard organizational units

- Use ridgelines, streams and ponds to start your patterns.
- Form ponds with buffer strips, plant trees along a fence line or road and work with contours on steep slopes.
- Build out from key features because they are important, unchangeable and a designated existing pattern.
- However, consider if they indeed will work for your long-term goals. Some current patterns are worth changing. You can open an old fencerow, change the orientation of garden beds and even move a poorly placed outbuilding.
- Think it through, then act.

60% planning, 40% action

If a standard unit doesn't fit, make it a uniform fraction

- We have an odd space that we cannot fit full 300' beds. We have modified it to be 150' beds and use it for productions that can be managed uniquely: hoop houses, nursery beds and early production.

Flexible equipment, tools and records

- As much as possible, we want our tools and equipment to be flexible between productions.
- We want records of yields to be equivalent to a standard spatial unit.

DESIGN TIP

Clearly lay out any ambitions now, so you can proceed slowly, knowing you've allocated space so you won't hamper future development of an area. Don't build a barn in a way that prevents you from digging into your hill for a root cellar. We planned our root cellar three years before we built it.

- We want to be able to adapt techniques throughout the property.
- We can use our flail mower to mow the edge of our caterpillar tunnel because we have designated a single organizational unit on either side as a *clover mow strip*. Then we mow along the edge of an electric fence in a pasture, under the orchard trees and along a linear sugarbush going up our laneway.

Thinking of the land as the equivalents of a standard raised bed is a fundamental breakthrough for the market gardener that is so approachable.

DESIGN TIP

Consider the fact that standardized tools, equipment and gardening techniques all enhance the ease of management. Easy management that allows greater and more diversified production on a farm.

Leave space using standard units

- As you develop your property, focus on priority areas.
- However, the space between these areas should be divisible by the standard units so they can be integrated more fully under management in the future.

Laneway Farming: An Example of Organizational Land Patterning

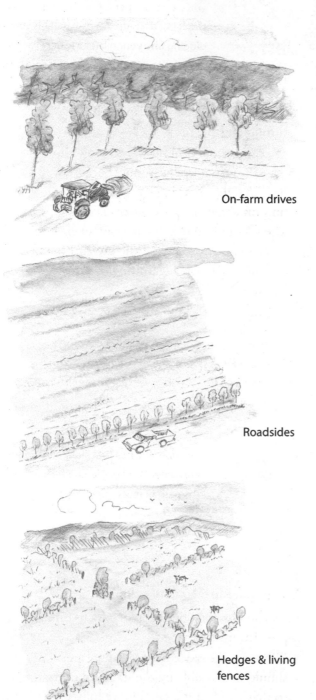

On-farm drives

Roadsides

Hedges & living fences

Do You Know?

Humans have been organizing land for many years. Maybe you have sat on a stone wall built from picked stones. Maybe you have driven along a maple-lined drive or noticed the patchwork of greens and browns and yellow from an airplane.

- For instance, we have developed using our standard unit moving out from key features: our ridge road, our lowland gardens and our fencelines. In between there are some undeveloped spaces, but we have included them into the patterns so eventually every square foot of our farm can be organized and patterned.

Standardize management for organizational units

- Once you commit to organizational land patterning, you can standardize your management approach based on a single unit, practical width for tools, or tractor, or equipment, or technique, etc.
We standardize on a 6' width; everything is in multiples of 6'.
  Examples:
  - Mowing is a standard way we manage edges through the farm; along a drive, along the edge of a garden, along the edge of a greenhouse, within our reforestation project. ⇥ 5.5' mower
  - Weed barrier is another standard we use. We lay it along edges of caterpillar tunnels, along edges of gardens to open more land, between nursery rows, as a ground cover for potting areas. We use 6' and 12' weed barrier.
  - Micro-sprinklers are a standard way we water. We move them over nursery beds

to germinate perennials, use them in the garden to water carrots, bring them into hot tunnels to establish a cover crop. We use sprinklers with a 18' or 32' coverage.
o All of this is based on our 6'x300' organizational unit.

Patterning

• Once you have your property or garden areas organized with standard units, you can pattern them.

• Patterning is done by assigning zones, alternating production places and grouping for guilds.

  For example:
o This is the garden area, that is the hedgerow and here is the pasture
o Every 12th garden bed will go into a tree species
o The beds adjacent will be planted into complimentary companion perennials.

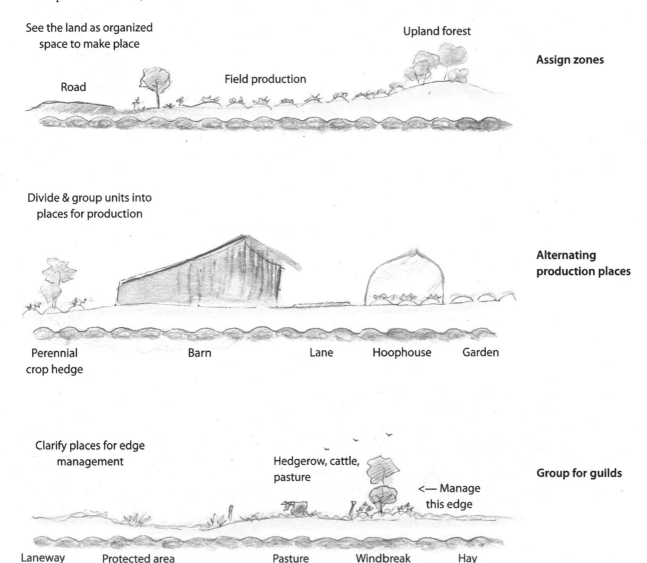

See the land as organized space to make place

Upland forest

Assign zones

Road

Field production

Divide & group units into places for production

Alternating production places

Perennial crop hedge

Barn

Lane

Hoophouse

Garden

Clarify places for edge management

Hedgerow, cattle, pasture

<— Manage this edge

Group for guilds

Laneway

Protected area

Pasture

Windbreak

Hay

Agro-ecological Land Patterning

Agro-ecological land patterning is focused on maximizing organized places for layered diversity of production. This could include alley cropping of cash crops, silvopasture, woodland market gardening, etc. It is land patterning for maximum integration of production. It considers:

1. The organizational zoning for farm management

2. The organizational units that are patterned for management of long-term production

3. The climate, ecology and human community which is paramount for production choices

4. It organizes complements of useable species that are ecologically and economically fit

5. It integrates undeveloped space into patterns for future patterning

6. It makes use of design management so it can evolve as an ecosystem yet still remain profitable like a farm.

Ecologically and Economically Fit

If a species is not able to grow well in your area or adaptable to environmental change, it is unfit. If a species is undesirable by the community or cannot be farmed profitably, then it is considered economically unfit. We also need to shift our perspective of profitability. Many species are profitable on a different time scale or provide essential services that are undervalued. For instance, oak hedges are affordable to establish. It could cost you as little as $100 to establish a fast-growing 300-foot oak hedge. In the market garden, the value may be sheltering your garden first, and then long-term nut and lumber production.

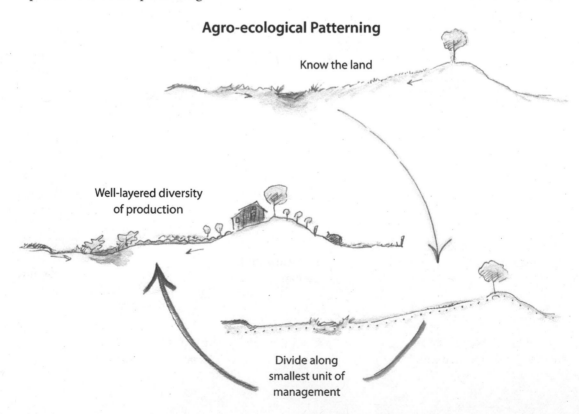

Agro-ecological Patterning

Know the land

Well-layered diversity of production

Divide along smallest unit of management

Smallest Unit for Management

You don't Need to Change the Space

Just because we are dividing the space doesn't mean you need work up every piece of soil or change it in any way. It just means you are recognizing the underlying grid of your land. It means that those places you do plow, build on or fence are in line with the rest of the property. This allows access, management, expansion and integration opportunities.

DESIGN TIP

With patterning the emphasis is put on the smallest common unit of management. This unit becomes the divider of space. I use the width of my tractor (6'). For a larger farm, this could be the width of their standard equipment (say 16'); for a smaller market garden, it may be only 3 or 4 feet.

Now and Future Place

Once land is organized, it is more easily mappable for present and future ecological patterns. Now we can conceptualize each unit as no longer space, but actually place. It has become a distinct place in space; in our case with a 6' width by a given length. Once established as a place, we can learn about its localized geology, pedology, hydrology and ecology and assign it a production goal. Say the soil is well-suited to a row of walnuts or pecans to grow up along your cattle pasture. Or perhaps a row of apples and pears beside your garden. Maybe you need to leave this space as an access lane, or the place is already occupied by a building.

How Many Organizing Units Does This Occupy?

Ask this question when approaching infrastructure, roads, hedges, and other more or less permanent units that already occupy space. How many organizing units does this occupy? You want to integrate the current ecology and humanscape of a property into your patterning.

Create a Pattern for Integrating Diversity

You can build any pattern you like. Alternating rows of trees and pasture, hedges and fields, garden beds and field houses. The more specific your pattern, then the easier to manage. For example:

NOTE: slash (/) delineates the edge between organizational units.

PR is pasture road, CL is clover, FH is fruit hedge, HH is hoop house, GA is garden. Organizational unit designations equal 6' (for instance PR/PR/PR is 18 feet).

- EX: /PR/PR/PR/CL/FH/CL/HH/HH/HH/CL/GA/GA/GA/GA/GA

You can conceptually pattern using:

- Word documents, slashes and acronyms
- Excel files with rows and columns and colors
- Templates made for the purpose

Spaces in Place

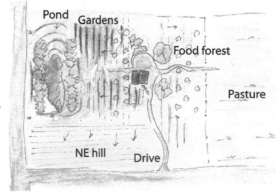

Intentional places in space

Organizational land patterning

The layers of the land

What to Consider When Patterning

1. What is the underlying nature of the land itself: soil, elevation, etc.?
2. What key features or patterns already exist: hedgerows, fence lines, roads, streams, hills, ridge?
3. What is the permanence of the features?
4. What is their current orientation relative to property lines and directions and other larger community ecologies?
5. What is the smallest divisible unit for design (e.g., the width of your tractor)?
6. See this unit repeating itself across the land; with repetition comes the ability to pattern.
8. Which way shall we divide: north-south, east-west, perpendicular to the slope?
9. Work with contours or with field edges.

Permabed System

Our system of permanent raised beds gives a framework for patterning. But it extends beyond the garden as invisible organizational units for managing the space between garden, hedgerows, pastures, orchards, buildings, etc. (see Permabed Section)

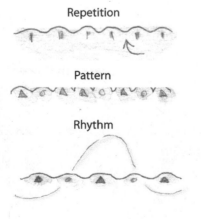

Repetition	Build beds
Pattern	Intercrop layer
Rhythm	Multilayer guild crop production

DESIGN TIP

Repetition allows patterning

Because you have many equal units repeated across a property, you can differentiate between these units through patterning. You can maintain 9 units for garden, 3 units as a linear sugar bush, 3 units as your road, 12 units as a wild windbreak, etc.

Scale of Permanence

When mapping your land for design zones consider the layers of permanence. Some things are easier to change. Consider the following:

1. Natural Permanent
   - You cannot change the sun, and it's hard work moving a hill.
2. Natural Changeable
   - You can cut trees down to change shadows or improve water retention in the soil.
3. Human Permanent
   - Your municipal road
4. Human Changeable
   - A fence
5. Flows
   - Most natural flows are permanent on a greater scale, the wind and rain and sun. But you can plant a windbreak, build a catchment pond and plant for shade.
6. Your Design Zones and Projects
   - These you have control over so site them so they benefit from more permanent layers, and avoid placement where you will have problems with these later (because they aren't changing anytime soon).

You can always amend the soil or cut a tree down, but you can't easily move a mountain.

Brainstorm of Farm Ecosystem Flows

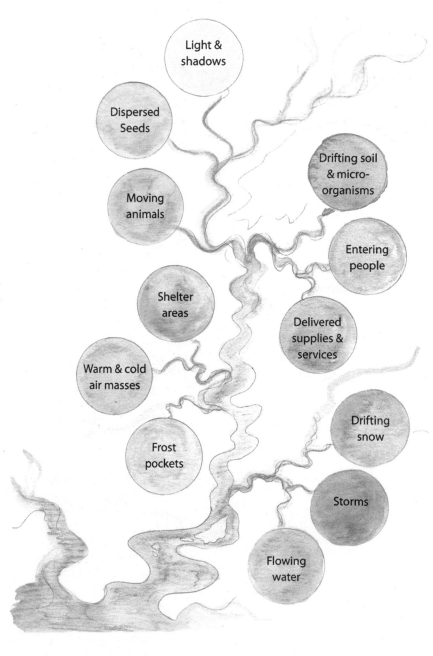

Mapping the Flows That Move Through Your Farm

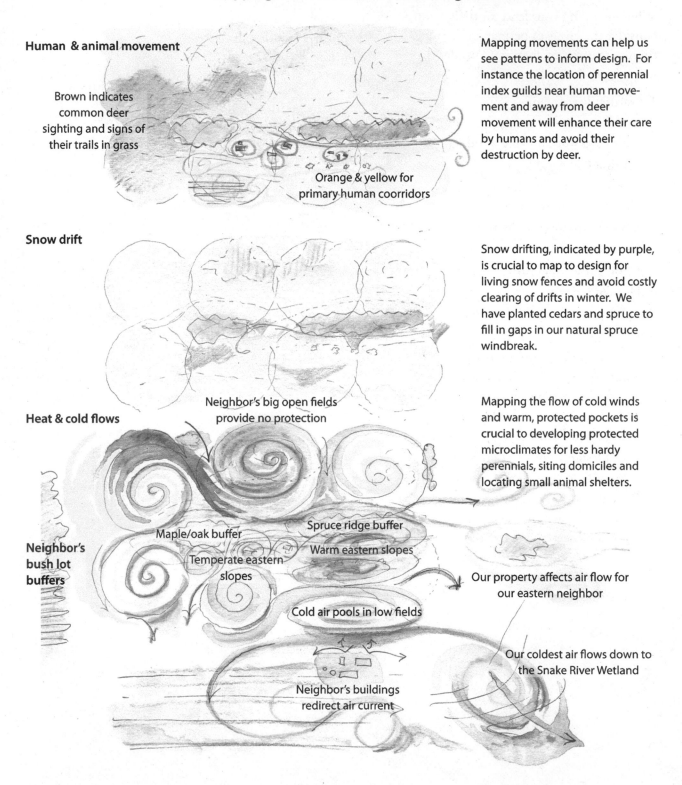

Human & animal movement

Brown indicates common deer sighting and signs of their trails in grass

Orange & yellow for primary human coorridors

Mapping movements can help us see patterns to inform design. For instance the location of perennial index guilds near human movement and away from deer movement will enhance their care by humans and avoid their destruction by deer.

Snow drift

Snow drifting, indicated by purple, is crucial to map to design for living snow fences and avoid costly clearing of drifts in winter. We have planted cedars and spruce to fill in gaps in our natural spruce windbreak.

Neighbor's big open fields provide no protection

Heat & cold flows

Mapping the flow of cold winds and warm, protected pockets is crucial to developing protected microclimates for less hardy perennials, siting domiciles and locating small animal shelters.

Maple/oak buffer

Spruce ridge buffer

Warm eastern slopes

Neighbor's bush lot buffers

Temperate eastern slopes

Our property affects air flow for our eastern neighbor

Cold air pools in low fields

Our coldest air flows down to the Snake River Wetland

Neighbor's buildings redirect air current

Design Management Zones

Design management zones (DMZs) are themed areas you allocate in your property. It is here that you take spaces and their distinct ecologies and microclimates and turn them into design areas for your future projects and productions. Whereas organizational land patterning is the underlying grid for linking the whole farm, DMZs are distinct places, microcosms within the property, nodes of activity: your barn and vicinity, for instance, or your nursery, home, gardens are places of intensive design and management.

Design Management Assessment and DMZ Mapping

Here is the process we used to map our land and establish DMZs, which are the foundation of how we operate (see discussion on the following pages for more details that pertain to this concept diagram.)

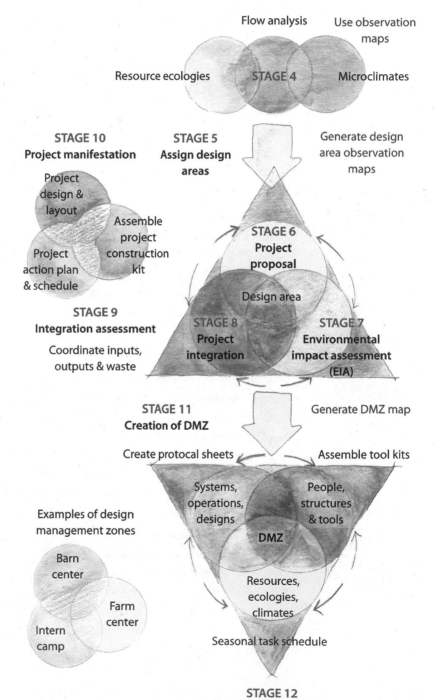

STAGE 1
Climate & Sector Analysis

STAGE 2
Property Circle Mapping

STAGE 3
Understanding the 12 layers of the land

Flow analysis   Use observation maps

Resource ecologies   **STAGE 4**   Microclimates

STAGE 10   **STAGE 5**   Generate design
Project manifestation **Assign design**   area observation
                          areas   maps

Project design & layout

Assemble project construction kit

Project action plan & schedule

STAGE 6
Project proposal

Design area

STAGE 9
Integration assessment

Coordinate inputs, outputs & waste

STAGE 8
Project integration

STAGE 7
Environmental impact assessment (EIA)

STAGE 11
Creation of DMZ

Generate DMZ map

Create protocal sheets   Assemble tool kits

Systems, operations, designs

People, structures & tools

DMZ

Examples of design management zones

Barn center

Farm center

Intern camp

Resources, ecologies, climates

Seasonal task schedule

STAGE 12
Whole farm design management

1. Climate and Sector Analysis

- Your first step is to know your climate, your hardiness zone.
- The greater flows of nature don't start or stop at your fence line, they are flows of regional and global scale.
- Place a point in the middle of your farm and use your protractor to make a circle around your property.
- Shade in areas with colors appropriate to major environmental flows into the property—animals, humans, rains, snow, wind, sun, etc.

Do You Know? Hardiness zones are areas designated in part by average minimum temperatures. Crops are said to be hardy to a zone's average winter low. However, this classification doesn't account for important factors like cold snaps, snow cover, summer heat, moisture distribution and the number of frost days, all of which can change winter survival.

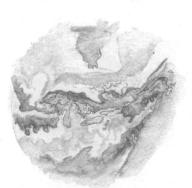

Take snow cover for instance. A crop that can withstand -25° C with 2 feet of snow cover may not survive when a cold snap comes early in a dry winter!

2. Property Circle Mapping

- Make a base map (print several copies).
- You can print a county map and work on top of this or draw your own.
- Divide your base map into equal circles using a sensible size relative to your property (10, 50,100, 1,000 acres) and the ecologies (distinct forest, meadow, plowed land) that exist.

A good view, from an office, wash area, etc., is an important design element — it rejuvenates the farmer.

Do You Know? Our climates are greatly determined by the distribution of heat and moisture by global atmospheric and oceanic circulation. Warm moist air in the tropics is moved north, and areas of high pressure occur over our drier regions. Our predominant winds are determined at this scale.

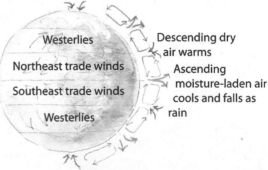

Westerlies

Northeast trade winds

Southeast trade winds

Westerlies

Descending dry air warms

Ascending moisture-laden air cools and falls as rain

3. Understand the 12 Layers of Your Land

- It is not necessary to map each layer individually, but you may wish to.
- It is more important to understand the layers that are available and incorporate or summarize this information within Step 4.
- Some maps are most useful:
  - Soils (soil maps are available)
  - Brittleness, water and macro vegetation (can be seen through aerial photography and assessed on-site)
  - Infrastructure, evident in most maps

4. Observe and Assess Zonal Differences: Resource Ecologies, Microclimates and Flow Analysis

- Use seasonal observation maps and a design notebook to move about your farm and formulate distinct zonal differences.
- *Resource ecologies* are areas of distinct resource potential for farm design of production areas as well as for infrastructure placement and other projects. We want to maintain them as ecologies, so they can continue to provide their resources and services.

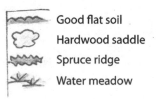

Good flat soil
Hardwood saddle
Spruce ridge
Water meadow

- Give these resource ecologies names on your property design base map. At this point, I name them as I see them. Use descriptive language and consider future uses. Use symbols on your map to help distinguish them.
  - Water meadow ➜ pond development
  - Wooded area ➜ wood chips, ephemerals, mushrooms, firewood, etc.
  - NE loam hillside ➜ orchard, early spring gardens
  - High-load capacity moraine ridge ➜ infrastructure, roads

- **Microclimates** are areas with climate differences, often created by exposures or protections from wind or sun. These could result in as much as a .5 zone difference within a small property. By analyzing microclimates, you are further defining your resource ecologies.

  Finding microclimates requires:
  - Flow analysis of wind, water, heat and frost
  - Windbreak analysis of trees, ridges and buildings
  - Obstruction analysis of buildings, roads, trees (these can cause frost blockage and pooling)

Cold open water meadow

Warm eastern hardwood saddle

Cool sheltered evergreen ridge

Temperate open southern field

Circle Mapping

LEGEND

Divide your base map into circles to help organize observations into zones. Circles are important (as opposed to squares) because they lend themselves to spherical observation.

Circular zoning helps understand microclimates and resource ecologies, since they suggest flow between them.

ROAD

Remember: your neighbors' properties interact with yours. Flow doesn't stop at the property line!

Map Your Elevations & Make Observations

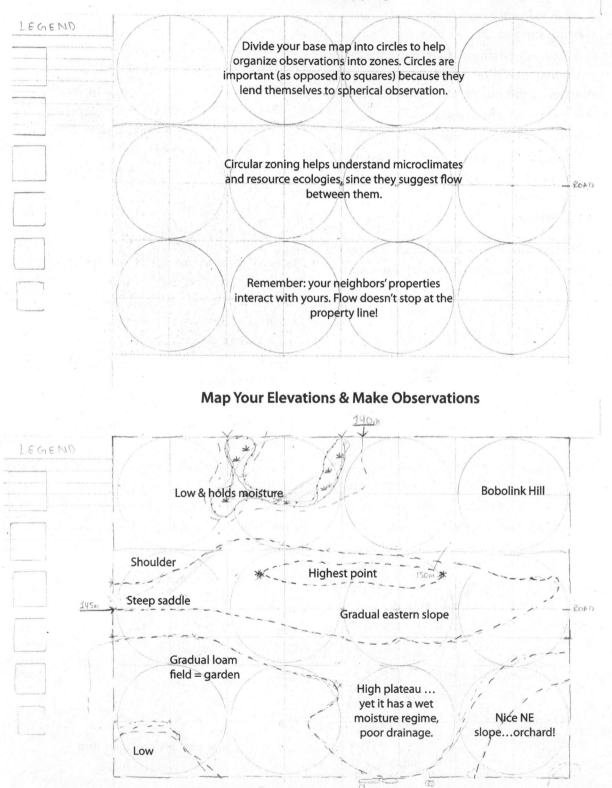

LEGEND

140m

Low & holds moisture

Bobolink Hill

Shoulder

Highest point   150m

145m   Steep saddle

Gradual eastern slope

ROAD

Gradual loam field = garden

High plateau ... yet it has a wet moisture regime, poor drainage.

Nice NE slope...orchard!

Low

Name Your Macro Ecologies and Make Them Stick!

Add your notes on their resources and microclimates

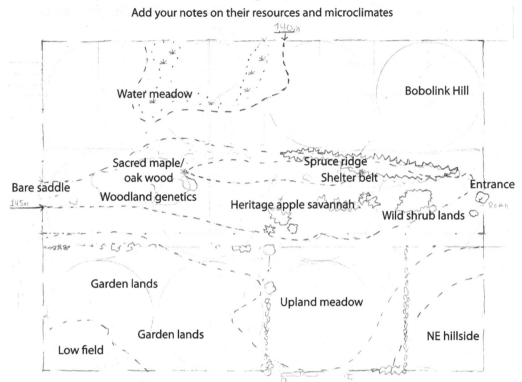

Water meadow

Bobolink Hill

Sacred maple/ oak wood

Spruce ridge
Shelter belt

Bare saddle

Woodland genetics

Heritage apple savannah

Entrance

Wild shrub lands

Garden lands

Upland meadow

Garden lands

Low field

NE hillside

DESIGN TIP

Naming your resource ecologies is another way of turning space into place, building familiarity for management.

Mapping Dry and Wet: Brittleness

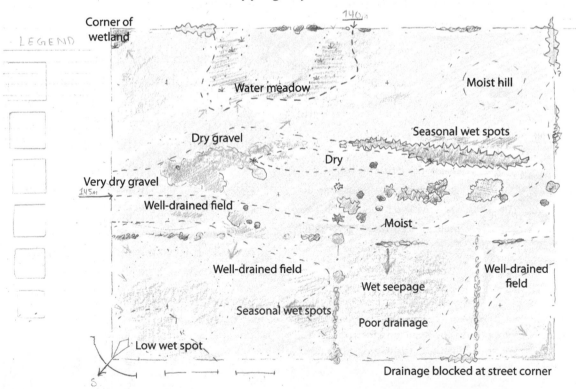

LEGEND

Corner of wetland

Water meadow

Moist hill

Dry gravel

Seasonal wet spots

Dry

Very dry gravel

Well-drained field

Moist

Well-drained field

Wet seepage

Well-drained field

Seasonal wet spots

Poor drainage

Low wet spot

Drainage blocked at street corner

Add to the names of your resource ecologies where they overlap with distinct microclimates.

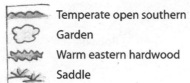

Temperate open southern
Garden
Warm eastern hardwood
Saddle

- **Human Flow Analysis**
  - This is assessing how people move through the farm and in and out of the ecologies and microclimates.
  - How accessible are the resource ecologies and microclimates? If they became design areas, could they be accessed for routine maintenance and operations?
  - Do a flow sketch on a base map of possible human movement. Use a symbol (*) to mark selected ecologies and microclimates, and find efficient ways across the physical landscape to get there. We are looking for sensible, low-cost road, lane and path planning.

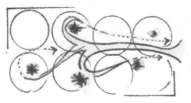

Size denotes importance to overall farm plan

DESIGN TIP

I have found that the best ways to get from A to B happen to be the ways marked through habit. I now build farm paths along trodden grass ways made before we had a way to move along.

5. Assign Design Areas to Your Property Base Map and Generate Specific Design Area Maps

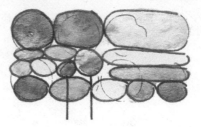

- The next step is official designation of some key design areas.
- These include your farm center and key infrastructure projects, and the roads and lanes that connect them.
- These also include primary garden areas and animal paddocks, irrigations ponds, perennial nurseries, etc.
- You can zoom in on these design areas and give them their own maps to help you better analyze the ecologies, microclimates and flows within them. You can make copies of these design area maps to help you sketch out potential projects.

6. Make Draft Project Proposal

Design areas will have several different projects, and every good project needs a proposal.

- Why is the project needed?
- How will it work with its ecology?
- What is needed to create it?
- How will it be integrated with other projects in the design area?

NOTE: A design area should have 3 proposed projects, potential productions that can work together to better use space,

A design area is a space on the farm that has been designated for a future DMZ, and it will include future projects, production and constructions, for instance, the site we designated to become our future garlic processing barn, root cellar and washing/packing area.

enhance time efficiencies and manage energy productivity.

Our Guild Enterprise Production

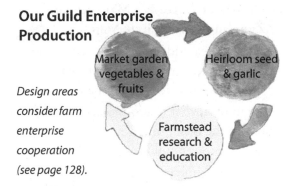

Design areas consider farm enterprise cooperation (see page 128).

Market garden vegetables & fruits

Heirloom seed & garlic

Farmstead research & education

Design Tips for Space/Time/Energy Productivity

- Multifunctional spaces
- Convenient work scheduling within the area
- Waste cycling as an input for production

Design Areas will evolve clusters of projects or project guilds. Examples include:

- Gravity irrigation/maple bush regeneration/hardy fruit nursery
- Intern cabins/shiitake mushrooms/mini personal gardens
- Wash station/curing shed/root cellar

Remember: you need not develop all your ideas at once, indeed, it is not recommended.

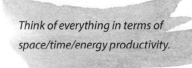

Think of everything in terms of space/time/energy productivity.

First, we should visualize the fullness of each design area and their connectivity across your property to better organize for future management.

You can focus on key projects for the property and develop these projects in relation to the overall design areas, other proposed construction or future production.

7. Environmental Impact Assessment (EIA) for Design Areas

- Analyze environmental sensitivity of design areas.
  - All environmental elements are assessed and mapped on design area maps (water flows, specific vegetations, sun, shade, etc.)
- Assess impact of proposed project.
  - How does a building change water or air movement?
  - How will a planting affect the current ecology?
- Identify areas of concern for sensitive ecologies. This can be minor, medium or major. (We have a small micro-ecology of sphagnum moss in our spruce ridge. I identified it to preserve simply because it is rare on my property)
  - We were careful to avoid destroying heritage apples and maples in the placement of our major buildings.
- We pay particular attention to water resources and avoid contamination and work to sustain levels reserves in our aquifer, soil and vegetation.

A project can be anything you build, grow or process. It could be a barn, a garden plot, an irrigation line or composting system.

- Design EIA concerns into projects and reform proposals.
  - Make note of any concerns, and address them through modifications in your proposals.
- Consolidate final project proposal

8. Integrate Projects into a Project Guild

- Lay out the three approved projects on your design area map.
- Integrate project inputs, outputs and waste with other projects.
- Our aim is to maximize using project waste and design for inter-project services.
- In our barn center DMZ, the compost from washing and packing is sent to a hugelkultur nursery for the production of perennials. The hugelkultur shades the wash station, keeping it cool for better vegetable processing conditions. Wastewater feeds the nursery, furthering vegetative growth and enhancing the evaporative cooling.

9. Integration Assessment

- Projects should form a guild.
- You must assess their interrelationships.
- Projects must be designed for integration:
  - Cycling waste
  - Improving layout
  - Provision services

10. Core Projects are Approved, Begin Project Manifestation

- Projects may manifest at their own pace.
- But at this point at least three core projects for current and future development should be approved for the design area to become a design management zone.

- Remember a project can be as simple as hugelkultur compost.

11. Creation of a Design Management Zone and DMZ Circle Map

- Now the area undergoes a key transformation into a recognizable place in space, a zone of ongoing design and management.
- Design of further micro-projects or additions to the core projects are possible, but the core project guild is the hub.
- Now is the time for inter-DMZ flow analysis and property integration by linking DMZs through organizational land patterning.
- The design area is now a DMZ and is referred to by a more telling name: the barn center, for instance.
- Take a property base map and periodically reorganize it as you develop DMZs.
- Use different sized circles and color codes to distinguish major and minor DMZs.

12. Whole Farm Design Management

WFDM considers DMZs as the spatial manifestation of your farm's guild enterprise production (see the next section).

- Eventually the inputs and outputs of all DMZs can be coordinated.

DMZs work together

A Farm Center/ Greenhouse

B Barn/Root cellar

C Gardens/Fields

- First look for solutions within each DMZ to make use of waste or for multiple uses of tools and human energy (through scheduling).
- Revise your human flow analysis and refine the roads, lanes, paths and alleys that connect your DMZ.
- Assign importance to DMZ as relates to your holistic goal.
- Build protocol sheets, task schedules and tool kits for each DMZ.
- Build a whole farm seasonal operation schedule.
  - This is a coordination of greater property management timing to avoid excess movement of people and tools around the farm when working in season.
  - Assign a permaculture property zone (1–6) to each DMZ, consider its accessibility and proximity to regular human movement for improved maintenance and operation management coordination with other farm activities and routines.

DESIGN TIP

In whole farm design management, we manage through our DMZs, their different processes, projects and productions, as well as using them as focus points for trialing and improving designs for better overall farm productivity.

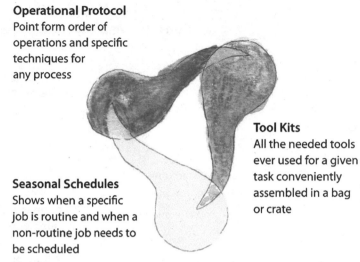

Operational Protocol
Point form order of operations and specific techniques for any process

Tool Kits
All the needed tools ever used for a given task conveniently assembled in a bag or crate

Seasonal Schedules
Shows when a specific job is routine and when a non-routine job needs to be scheduled

For instance:
- We do big harvests every Wednesday and Friday
- We frequently sow cover crops in May/June and again in September/October
- We mulch our garlic 1st week of November

Primary, Secondary and Tertiary Design Management Zones

As your property evolves with many new projects and increased productions, it becomes possible to look down at your farm from above and see your DMZs relative to each other. Some are critical to your core enterprises, and some are peripheral. Select 3 primary, 3 secondary and 3 tertiary. This will further focus your management priorities. When starting a farm, focus on your 3 primary enterprises. For us these were: our farm center, barn center and garden.

We also managed our laneway, research nursery and reforestation area.

When Mapping for Design Zones
Consider Map Layers in Order of Permanance

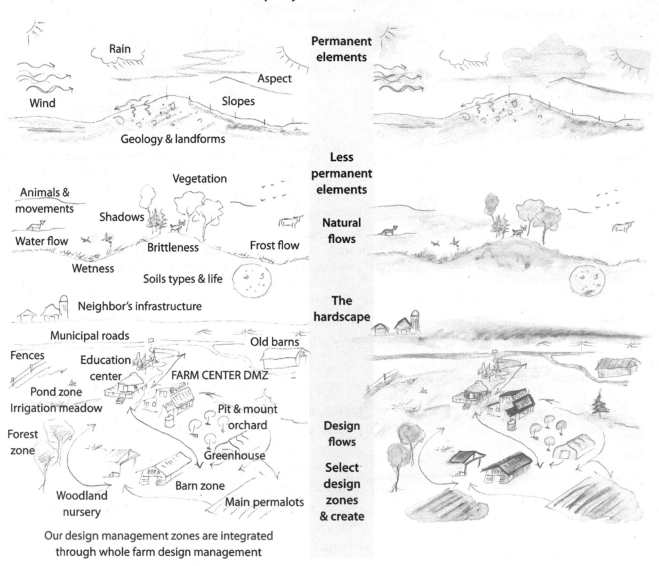

Permanent elements

Rain
Aspect
Wind
Slopes
Geology & landforms

Less permanent elements

Natural flows

Vegetation
Animals & movements
Shadows
Water flow
Brittleness
Frost flow
Wetness
Soils types & life

The hardscape

Neighbor's infrastructure
Municipal roads
Old barns
Fences
Education center
FARM CENTER DMZ
Pond zone
Irrigation meadow
Pit & mount orchard
Forest zone
Greenhouse
Barn zone
Woodland nursery
Main permalots

Design flows

Select design zones & create

Our design management zones are integrated through whole farm design management

Farm Design and Management Terminology

Although these can be used interchangeably, to some extent, depending on the individual's experience, I think it helps to put them on a graduating scale to develop the concept of design management.

Farm design is the placement of project, infrastructure and production areas within a property. Emphasis may be put on the best structures for the production and improved organization of productions.

Whole farm design is the placement of these in relation to the land itself (its ecology) and in consideration of each other. It often emphasizes resilience and regenerative systems and may consider a holistic management plan (such as HRM, page 111).

Whole farm design management considers the above definitions which, although talked about at length in another section, require some preamble now, and is uniquely linked to principles for design management (see page 143). It also emphasizes designed production such as guild enterprise production (see page 127), which coordinates beneficial exchanges between diverse commercial productions. It is a focused model for viable diversified farm businesses matched to a property, community and farmers. For instance, we have three main businesses: heirloom seed, fruit and vegetables, and education and research. These three enterprises form our guild enterprise production, and whole farm design management in a framework for integrating the processes, infrastructure and productions across the property. It focuses on building the inter-relationships needed to manage three core enterprises successfully while maximizing the land's natural resources.

The following contribute to whole farm design management:

Organizational land patterning is the underlying grid of the landscape that we design upon following pattern-ready and uniform organizational units.

Agro-ecological land patterning is the specific use of smaller organization units (although not necessitating more intensive land management), to differentiate space into unique places for the integration and management of agro-ecological diversity.

Design management zones are discernable areas of at least 3 key projects, infrastructure and production whose processes are managed for mutual benefit, both within the DMZ and between DMZs.

Principle

Rule of Three

We consider three as a unifying design principle because design should have both a motivation for more diversity and a limit to it. Three works well because it is the first number where there are complex interactions, the first number where true community exists.

Permaculture property zones assess the overall accessibility and frequency of visitation to areas on the farm (DMZs). This helps inform how we manage areas and how to develop projects based on how much attention they require throughout the season. Some projects, like our wild leek production in our forest, only require two visits a year. First we go to forecast the yield quantity and maturity timing, and again to selectively harvest. This is a zone 5. Our farm centre is zone 1.

Permaculture zones help us understand how frequency of movement into an area affects its best management

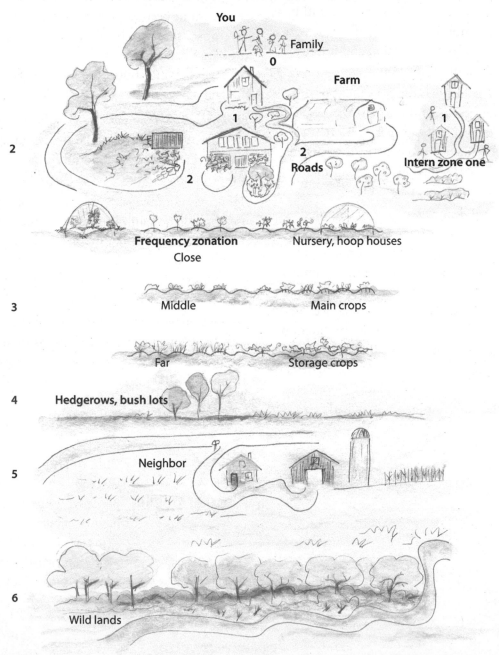

Farm Permaculture Zones

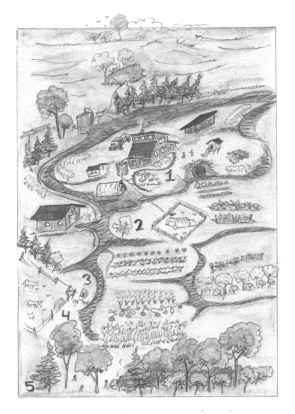

Our Farm Centre Design Zone

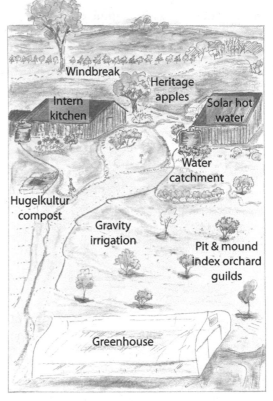

This area is the most trafficked zone. It is the connective tissue between living and working.

Index guilds can be planted in hugelkultur compost.

Root Cellar & Barn Design Management Zone

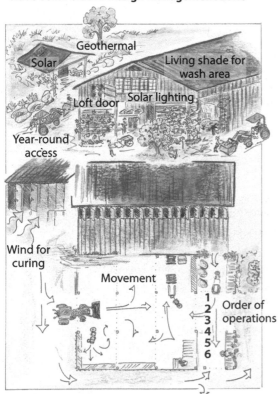

Our Intern Camp

This has been an evolving design. Our interns in our first years lived in tents. When we moved to our current farm, we selected this quaint, sheltered circle of spruce for them. Here is the final design we work on each year.

Mapping Tips

1. Make full use of the graphic elements available: colors, symbols, names, shades, etc.

2. Say it as you see it, use descriptive place names: Wet Spot, Moist Hill become cold low-lying water meadow and moist, sunny SE hillock.

Proper garden character mapping can make all the difference in early production, perennial site selection and soil-specific cropping.

3. Use existing maps for data like elevations, soils and macro-ecologies.

4. Make sure to compare the data found on regional and local maps with field observation, testing and further research. This is important for critical resource analysis. For instance, soil type is pretty important to a market gardener, but my county soil map isn't as exact as my needs. For instance, this primary garden field of ours is all classified as one soil type, but in fact it has large variations in soil parent material, texture, stoniness, hydrology, etc.

Mapping Your Garden's Soil and Water
Garden Environment Mapping (GEM)

Principle

Some of the most important elements to map in your garden plots:

Geology: rockiness, soil texture, stoniness

Pedology: fertility, pH, soil life

Hydrology: drainage, springs

Ecology: plant species, animal movements

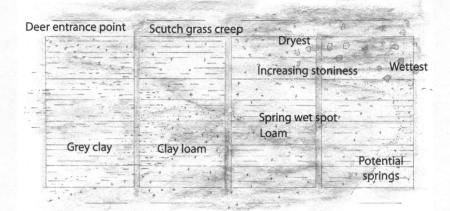

It can be surprising what you learn when you take the time to make quality observations. For instance, our dryest and wettest areas in the garden are immediately adjacent and both on the highest ground.

Observation can be done throughout the year, with some seasons giving different revelations. It is easier to track animal movements in winter and early spring. It is best to determine dryer areas for early production in spring when you can evaluate the workability of the land.

Some things are best monitored through multiple seasons for better understanding. For instance, a wet area needs to be monitored to see if it dries up in early summer or if it remains moist into fall.

Holistic Planning

Defining the Whole and Making a Holistic Goal

UNDERSTAND YOURSELF: IKIGAI

Ikigai Sketch

Before we can truly approach farm planning, we need to begin the journey by searching within ourselves, our reason for being. The Japanese concept of Ikigai is the best I have encountered for encapsulating what it means to find your true path. What aspects of farming attract you most? Which aspects are you best at? Which aspects are people willing to pay you to do? Which aspects do our communities and world need?

All decisions are made to move us toward our goal.

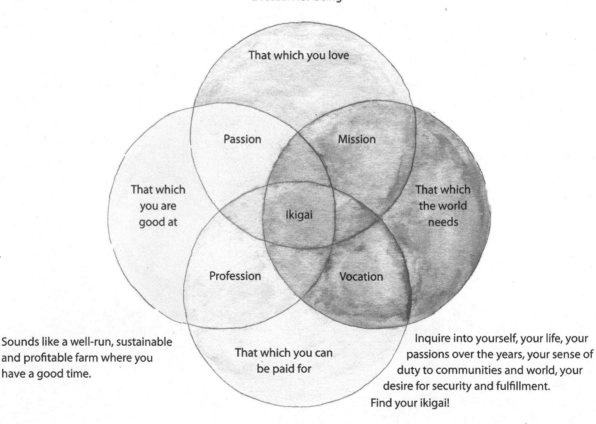

Ikigai

"a reason for being"

That which you love

Passion

Mission

That which you are good at

Ikigai

That which the world needs

Profession

Vocation

That which you can be paid for

Sounds like a well-run, sustainable and profitable farm where you have a good time.

Inquire into yourself, your life, your passions over the years, your sense of duty to communities and world, your desire for security and fulfillment. Find your ikigai!

Holistic Resource Management

In building a profitable farm we have drawn upon knowledge from many places. One of the beneficial strategies introduced to us by my father-in-law, a grass-fed beef farmer, is holistic resource management (HRM). HRM is a framework for seeing the interconnectedness and better planning the way we interact with our land, community and family.

Although this framework is most often used on livestock ranches, it has much to offer the market gardener. Please conduct further research about aspects we don't discuss. The ultimate take away from HRM for the purpose of this book is the creation of a holistic goal and the testing questions for decisionmaking.

Defining the Whole

First things first. HRM asks us to define the whole.

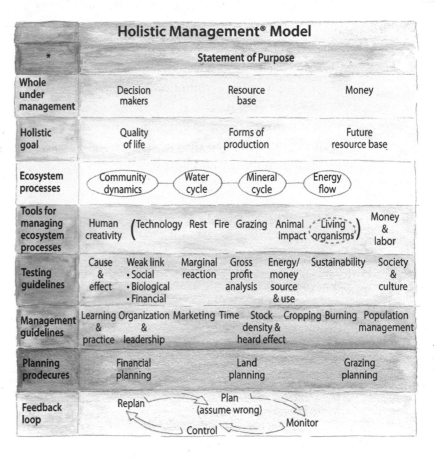

Holistic Management® Model

*	Statement of Purpose		
Whole under management	Decision makers	Resource base	Money
Holistic goal	Quality of life	Forms of production	Future resource base
Ecosystem processes	Community dynamics	Water cycle — Mineral cycle	Energy flow
Tools for managing ecosystem processes	Human creativity (Technology Rest Fire Grazing Animal impact Living organisms)		Money & labor
Testing guidelines	Cause & effect Weak link • Social • Biological • Financial Marginal reaction Gross profit analysis Energy/ money source & use Sustainability		Society & culture
Management guidelines	Learning & practice Organization & leadership Marketing Time Stock density & heard effect Cropping Burning		Population management
Planning prodecures	Financial planning	Land planning	Grazing planning
Feedback loop	Replan — Plan (assume wrong) — Monitor — Control		

Take a moment and define the whole of what you will be managing and that which is influenced by your decision-making.

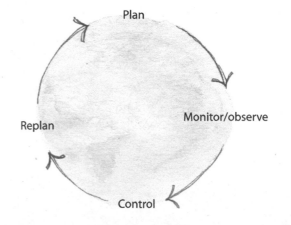

Resource base

These are all the resource of your land, buildings, equipment and community. This includes your customers, neighbors, suppliers and others who are involved in the well-being of the farm.

Decision-makers

Those involved in the day-to-day decision-making of the farm.

THE WHOLE

Money

This is important to consider separate from your resource base, because although you may have fertile fields, they are not as liquid as money in the bank. This includes cash, savings, investments, lines of credit, crowd funding & money from relatives.

Do You Know?

To get where you want to go, it is important to understand the *cycle of goal achievement*. 1) PLAN where you want to go, 2) MONITOR your progress, 3) CONTROL situations when obstacles occur, 4) and REPLAN to avoid them in the future and remain on purpose to our PLAN.

Planning Process

Plan

Monitor/observe

Control

Replan

Holistic Goal

Every farm should have a holistic goal, which helps determine a future that includes how you want to live economically, socially and environmentally. This is important because it informs decision-making, ties your values and productions to the environment and local community and puts your lifestyle and goals to the forefront.

We should farm the farm *and not let the* farm farm us.

This has three parts:
* Quality of life statement
* Forms of production
* Future resource base

Quality of Life Statement

Forms of Production

Future Resource Base

Why am I farming and how do I want my life to be?

State what you want broadly:
"We want to grow healthy food in a profitable and sustainable manner."
"We want to be healthy of body, mind and spirit."

Make clear statements:
"We want to improve our soil through compost, green manures and reduced tillage."
"We want to be able to travel each winter to visit family and friends"
or
"We want to take our kids camping in August."

It is so important to plan for a lifestyle you want, otherwise farming can quickly take over any other life desires.

What is the production needed from the land? Quality vegetables, hay mulches, trees and shrubs? What production is necessary to attain the life we wish to lead. What will you produce and how will you market it? This is the broad goal-oriented start of a farm plan.

Needed resources for continuing production.

The farm's behavior and perception by customers, family, suppliers, etc.
We cannot maintain a given production when we lose our image as honest, sustainable producers.

How we want the land to be:
"We want healthy soil, abundant water, diverse trees and habitat for beneficial organisms and native species."

How we want to see our future community:
"Our community is thriving with small businesses, good schools and natural places for learning and meditation…"

Holistic Goal

Focus your planning towards

Farming for your lifestyle

Facilitates

Decision-making

Connects production to the

Environment & community

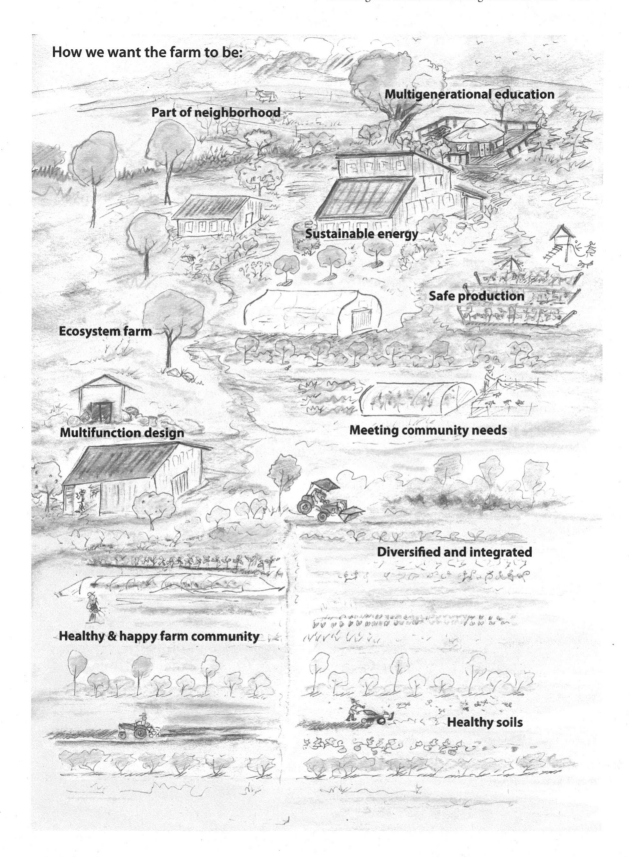

How we want the farm to be:

Multigenerational education

Part of neighborhood

Sustainable energy

Safe production

Ecosystem farm

Multifunction design

Meeting community needs

Diversified and integrated

Healthy & happy farm community

Healthy soils

PLAN FOR THE FUTURE

When you know where you want to go, you'll get there. Whether it is setting the objective of weeding your way down a row of carrots, setting intentions for seasonal goals or planning for your retirement. We must know where we want to go. Not only know it, but state it! Personally, clearly and positively state it!

- We will double our Community Supported Agriculture (CSA) income by 2018 through improved social media, more diversity and flexibility in our program and season extention to keep our customers supplied through winter.

- We will finish this bed of carrots in 30 minutes.

When planning, start with the future in mind.

By planning for a specific future production we can design for an efficient and profitable transition.

Here are some ideas for permaculture-style production for the future:

Medicinal herb wholesale

Fruit share CSA

Essentials oils production

Year round market garden

Specialty nursery

Specialty events

Pick your own forest garden

Small livestock and market garden

Mushroom production

Specialty crops for beer and wine

Native plant nursery

Rotational grazing with edible living fences

Approach your property with the question: "What do you want me producing?" Rather than: "Here is the productivity I want to impose on you."

- We will increase root cellar sales by 33% with new wholesale accounts for carrots and potatoes.

- We will set aside 10% every year for retirement.

8 Testing Questions

There are many important decisions to be made on a farm. Having a good framework for decision-making is especially important in the early years when everything is new. The holistic goal is a viewpoint for guiding all decisions, but you also need a sounding board for quick decisions. Holistic resource management uses a series of seven testing questions for decision-making (see next page).

DESIGN TIP

Plan by 3s

We make a 3-year, 6-year and 12-year plan. We also consider what we want to leave the land like to the next generation.

Do You Know?

Consider Potential Futures

Consider your land, use your observational maps and analyze potential design areas, remind yourself of your Ikigai and holistic goal. Now brainstorm potential futures on the land.

It is easy, when busy, not to make time to test your decisions. DO IT! Use these testing questions or something similar.

But use something to guide your decisions whether big-budget questions or field-management questions.

Testing Questions

Note: example questions for testing, discussion and answers are in *italics*

Root Cause

Does this action address the root cause of the problem? Should we buy a bigger delivery truck?

We want to be able to expand our CSA numbers at one of our drop locations, and currently we use our truck for two locations and cannot fit more on that day.

Weak Link

- *Social:* Are there any social concerns regarding this action? Will all farm hands be able to drive it? What training will be required?
- *Biological:* Does this action address the weakest point in the life cycle of this organism?

  This is not applicable for this question, because we are talking about a truck, not an organism.
- *Financial:* In my enterprise, what single thing will have the greatest positive impact on my ability to generate more income?

  We could also increase the value of baskets and not the quantity.

Comparing Options

Which action gets the "biggest bang for the buck" toward your holistic goal? Where is your highest return?

Reducing the drops the farm truck makes and using our car for those big weeks (melons and squash) frees up space at the critical time and so saves energy! We could rearrange our two drop locations so the big weeks (when biweekly and weekly shares pickup DON'T occur on the same week), for those drops that share our vehicles. This way we could increase shareholders at the drops.

Gross Profit Analysis

Which enterprises contribute most to cover the fixed costs (overhead) of the business?

This gives you perspective into where you should be investing based on the actual income the different enterprises provide. Our CSA is a very important income source, and a new vehicle would give us a much easier distribution process and enhance our perception by the community.

Input Analysis

Is the energy or money to be used in this action derived from the most appropriate source in terms of your holistic goal? Will the way the energy or money is to be used lead toward your holistic goal?

More vehicles for the same distribution does not meet our holistic goal of sustainable production and multiple functions. The big vehicle would get little use during the rest of the week, as it has way more space than necessary for our farmers market sales. We are also slowing market sales (trucking), in favor of increasing on-farm sales. THINK AHEAD before you BUY BIG!

Vision Analysis

Does this action lead toward or away from the vision articulated in your holistic goal?

Since our updated vision is toward not expanding our production further, but refining our current capacity to higher value production, then NO, increasing our vehicle capacity for more production is not following our holistic goal. When we look deeper, we see we want to actually shift where our income comes in more than expand production and so eliminate these bottlenecks of vehicles and time during busy season.

Gut Check

Considering all the testing questions and your holistic goal, how do you feel about this action or decision now?

Your GUT can always give you the answer: MY GUT says this is unnecessary at this time!

Common Decisions to Make

There are many important decisions to be made in farming. If you are just starting out, the list will go something like this:

- Why do you want to farm? Don't ignore this one. It deserves the majority of your attention.

Start with why!

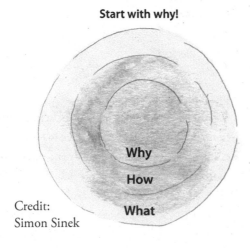

Why

How

What

Credit:
Simon Sinek

- Consider posing the following questions as why, how and what questions. Pay particular attention to the *why*.
- Where will you farm? Country, suburbs, urban center?
- Will you own, lease-to-own, rent, barter for, or incubate on land?
- Selection of potential enterprises: garden, chickens, orchard.
- Business model: partnership, cooperative, sole proprietor.

Know your skills and use them to save money. Are you good at construction, organization, mechanics or design?

- How will you lay out your property or parcel?
- Type of equipment scale to use: walk-behind tractor, small tractor, horse power, hand tools and carts.
- Infrastructure types and uses: barn, wash station, cabins.
- Filling your labor needs: self-run, volunteers, internship program, employees, etc.

Example questions for testing:

Should we expand to another market?

Should we hire an employee(s)?

Should we replace our old tractor with a new one or another used tractor?

Should we grow brussel sprouts?

PLANNING TIP

Can I be Profitable without Land?

Land is often seen as an impediment to successful market gardens. But this is far from the truth: It would only cost us $500-$1000 per year to rent the land needed for our business. It is skill, labor, product quality, timing and marketing strategies, etc., that impede successful farming more than land. Renting land, incubator farms, cooperative farming and affordable properties can all be found.

Do You Know?

Incubator farms host new farmers and their commercial ventures.

Perspective

When planning, designing and managing your farm, it is important to gain perspective. Literally! Climb a hill, charter a plane, lookout of the barn loft window and gain some perspective on your operation as a whole. Don't have tunnel vision, step out of the daily grind and see how things are unfolding.

You miss a lot when you have tunnel vision; have goals and pursue them, but open your eyes and commit to making a few mistakes every year so you know you are growing, stretching and innovating.

Community Resources and Peer Networking

It is important when planning to incorporate your community resources. It doesn't make sense to invest in a plow if every neighbor has one. It doesn't make sense to research the best tomato variety for hours until you first ask your peers. These resources include:

- Family and friends: support, advice, encouragement, sounding board
- Neighbors: local wisdom, access to equipment, manure and other resources, possible partnership

- Local community: demand for certain food stuff, encouragement, possible partnerships, local business solutions and resources
- Greater community: trends in your field, knowledge bank, conferences, workshops, online forums, magazines
- Peer farmers: advice, brainstorming, knowledge sharing, equipment discussion, crop favorites
- Leaders in your field: their experience and long standing examples.

Sometimes all you need is a cup of tea and a little perspective.

Holistic Budgeting for Profit Resilience

Profit Resilience

Profit resilience is the investment in eco-system services, including the investment in yourself and your community so that they can support you in return.

Investment in Ecosystem/ community/you

Helps production for efficiency, quality and better management

Improves your continued production.

Bring profit for you/community/ecosystem

Know How to Succeed

In order to reach profit resilience we need to know how to get there. We have already discussed many strategies to help you. Indeed this book is about agro-ecological profit resilience for farmers. You must understand the land, understand how to pattern, understand

who you are. But now you must budget. Budget for not only your seasonal production, but also for supporting ecosystem services!

Budgeting Holistically

To budget holistically, we must consider the farm business needs, your needs and the the functioning of the agro-ecology. What follows is a mix of the farm planning and budgeting strategies I learned from holistic resource management and the budgeting system born from my growing concerns on how to plan for profit resilience. Ensure the farm is actually building more soil, layering diversity and increasing production possibilities, while providing a good livelihood for the farm each season.

Profit First Budgeting

There is a major different between holistic financial planning and the more typical cash-flow model. HRM financial planning is very beneficial to the market garden because it is profit-oriented and organized around paying farmers a wage!

Let's Track the Differences

Holistic Financial Planning	Cash-flow
Profit is the goal	Production is the goal
Profit is planned initially	Profit is what is left over after expenses
Expenses are in categories	Uses overhead and variable costs
Monthly monitoring to stay on track	Often annual monitoring
*HRM expense categories (see page 120)	

Conventional cash-flow budgeting focuses on allotting money toward expenses (capital, variable and fixed) based on predicted income. The goal is to keep your costs below your gross by means of records (income and expense) and forecasting the trend for the coming year.

However, this doesn't always occur. Many factors can scoop up profit, especially when partially dependent on capricious nature. A crop failure can cut deep, or a month of cold/windy markets could reduce sales.

Many farmers don't pay themselves a wage, like any other employee. The year-end profit is their wage. Or what is left of it in a bad year.

Real Profit

Holistic financial planning, on the other hand, sets aside profit from the get-go and drastically changes how we arrange the remaining budget.

If we set aside salary for the farmer from the get-go, it can't be let go. It is unavailable for other expenses. That is the farmer's wage, because he will be working all year long and needs to be paid! Any money left at the end of the year after expenses will actually be surplus (or real profit). Real profit is the year-end monetary surplus after covering all costs of running the farm, including *all* labor.

In this model, the farmer can always make an income. As you grow your farm, your forecasting of income will be more precise. Records of market sales and CSA distributions is paramount to improving income projections. Consider your product value, assess efficiency, and remodel your business for better return.

Don't risk your wage on the weather, plan for profit first!

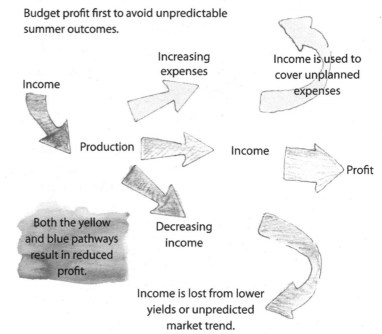

Budget profit first to avoid unpredictable summer outcomes.

Income → Production → Increasing expenses → Income is used to cover unplanned expenses

Production → Decreasing income → Income → Profit

Both the yellow and blue pathways result in reduced profit.

Income is lost from lower yields or unpredicted market trend.

Farms must budget for the farmer first. Consider this like an emergency airline announcement: "If the farm requires assistance, secure your oxygen mask first, and then assist the others." It doesn't matter what production dilemma occurs, you must take care of those in charge first. Otherwise it is too easy to have all your profit go to making production better and fighting fires, and not paying a living wage to you.

PLANNING TIP

When it comes to:
- tools; buy well-made
- specialized equipment; buy new
- equipment with few moving parts; buy used
- seldom-needed and common equipment (plow); pay a neighbor to do it

and always ask opinions of other farmers before buying anything.

You can meet your financial goals in a multitude of scales and situations. Be clear about what you need as your wage in your first year and third year and then your goal for your wage over the next 15 years.

Finding Better Return

If you are not making enough income running your farm, you must consider change. Perhaps running an urban CSA isn't as profitable as being a small plot grower. In the city, lettuces and niche salad mixes will return much higher profit per square foot. Since you cannot sustain a CSA on salad alone, perhaps you should consider restaurant and market sales, where you develop a specialty niche for products that can be sustained on a small acreage. Always consider your limiting factors. What is your weakest link?

On the other hand, in the countryside, if you are far from these wholesale accounts and you have space, consider a CSA model, where crops like squash, potatoes and peas, although space intensive, are less labor intensive and are great for CSA baskets that provide upfront reliable income.

Consider how you are limited by space, by time and by energy.

HRM Expense Categories

After planning for profit, the following budgeting categories: Wealth generating, Inescapable and Maintenance (WIM) are very helpful for organizing how money is allocated. This can make a big difference when we need to prioritize expenses to maintain our profit-first relationship with the budget. The categories include:

1. Wealth generating

These expenses go directly to producing profit for the farm.

- For us this would include row covers, composts, seeds
- All of these make immediate returns for the farm

2. Inescapable

These expenses must be paid.

- Taxes, land payments, loans

3. Maintenance

These are important to keep the business running fluidly.

- Equipment repair, truck oil changes

Monthly Monitoring

This is important because it ensures we catch any movement away from our budget, nipping profit leaks in the bud! This includes avoiding these tendencies:

- Allowing production costs to rise until they equal estimated income levels. These levels are usually optimistic, and such a tendency could steal profit if income is drastically lower.
- Borrowing too much in anticipation of optimistic income later in the season.
- Insufficient planning or planning for production instead of profit. Planning for production includes such common mistakes as filling empty garden space with more plants "just because you can." Oops, did you plan the labor needed to sustain these, or the marketing they'll need to move them?

Our budgeting is ever evolving, but having a framework has been essential.

BROWN BRAIN INVESTMENT

Over the years I have added to our budgeting framework to help us better invest in supporting ecosystem services for profit resilience. Brown brain investment is the culmination of this budgeting evolution. Perhaps it simply stemmed from justifying my purchases of trees to my wife. But really it comes down to knowing we need to invest annually in diversity, perennials and soil. Without a doubt, this is inspired by my father-in-law, who, when I first came to the farm over a decade ago, told me he had been planting 1,000 trees a year for the past 30 years.

I immediately began planting trees with him every spring on the family farm. Each year we put another 1,000 trees into the ground. He would send me off along a fence line with a bucketful of oaks or a bag of cedars. I planted upland shelterbelts, riparian buffers, windbreaks and living fences. Now, my favorite places on the farm are those I can go to and see those trees growing out of the ground.

Places Planted with a Brown Brain

- The walnut riparian strip by the big alfalfa field is now a stunning thicket of young nut trees.
- The elderberry plantation by the gravity-fed watering system has been proffering up fruits for years and we have now begun to intercrop other species in their shade, mulching them down with old elderberry stems.
- Our heritage sugar maple laneway.
- The diverse trees we are intercropping into our gardens.

Brown Brain

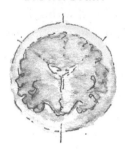

A new way of thinking

Brown Brain Investment

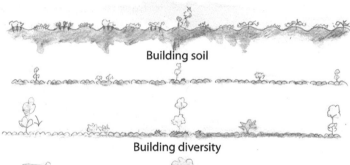

Building soil

Building diversity

Building profit through products and services

Building resilience

When I go to those places first planted 30 years ago, I am blown away by the richness.

- The willow-lined spring-fed solar pond that provides water pumped to a sheltered water trough
- The elderberry, cranberry and Siberian pea shrub that line a series of duck ponds that drop down to...
- The great reservoir, lined with cedar and spruce, butternut and ash, oak and elderberry

- Here wood duck boxes, turtle platforms and bat boxes are near.

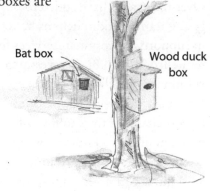

Bat box

Wood duck box

Layering Diversity in our Elderberry Planting

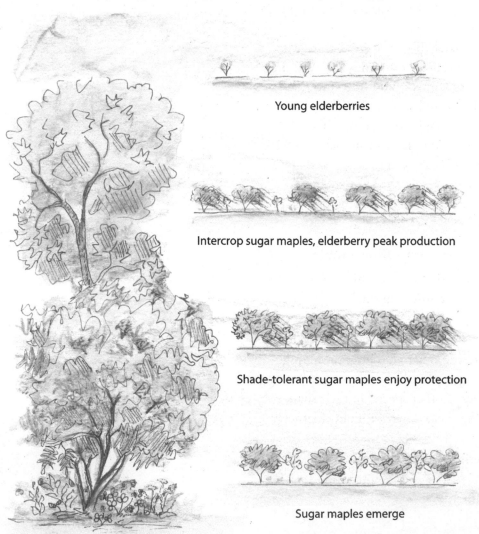

Young elderberries

Intercrop sugar maples, elderberry peak production

Shade-tolerant sugar maples enjoy protection

Sugar maples emerge

We Add Another Three Categories to Our Budgeting

Inspired by our experiences for a resilient farm, we expanded our budgeting techniques and now include brown brain investment.

1. Perennial percentage: 1% of gross income goes toward perennial plantings for three years, 2% for the next three years and 3% from then on.

- "We are committed to the continuous planting of trees for profit, pleasure and the greater good." (from our Holistic Goal.)

"There he began thrusting his iron rod into the earth, making a hole in which he planted an acorn; then he refilled the hole. He was planting oak trees. I asked him if the land belonged to him. He answered no. Did he know whose it was? He did not. He supposed it was community property, or perhaps belonged to people who cared nothing about it. He was not interested in finding out whose it was. He planted his hundred acorns with the greatest care."

— Jean Giono,
The Man Who Planted Trees

2. Future funds

Children's Future fund: We put aside 3, 6 and now 9% of farmer earnings toward a children's future fund and our own savings plans.

- Allocating income to future funds during the winter budgeting time ensures we set income goals that reflect long-term savings as well as a good living wage. Farmers cannot start saving for kids and retirement down the road, because that's when you need it.

3. The whole soil 30%: We budget 30% of our wealth-generating expenses to directly benefit soil improvement.

- This can include many expenses that farms incur anyway, but it ensures we never neglect soil improvement when deciding between different inputs.

- For instance, cover crop seed, manure, inoculants, composts, hay or straw mulches are all good uses of funds and they directly improve soil.

- Equipment, vehicles, bins, crates, and signage don't improve soil. However, if tools or equipment especially reduce tillage (for instance), consider 50% of its value for the whole soil 30%.

Children's Future fund is an annual allotment of income for the next generation and an essential component of brown brain investment. There cannot be a legacy of stewardship on the farm without the funds available to educate and train the next generation of small-farm entrepreneurs.

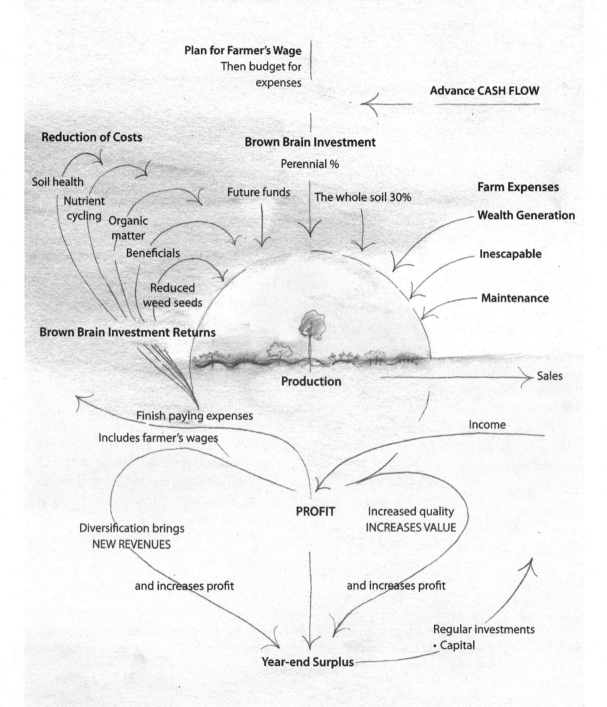

Profit-first Budgeting and Brown Brain Investment

Plan for Farmer's Wage
Then budget for expenses

Advance CASH FLOW

Reduction of Costs

Brown Brain Investment

Perennial %

Soil health

Nutrient cycling

Future funds

The whole soil 30%

Farm Expenses

Organic matter

Wealth Generation

Beneficials

Inescapable

Reduced weed seeds

Maintenance

Brown Brain Investment Returns

Production

Sales

Finish paying expenses

Includes farmer's wages

Income

Diversification brings
NEW REVENUES

PROFIT

Increased quality
INCREASES VALUE

and increases profit

and increases profit

Regular investments
• Capital

Year-end Surplus

CASE STUDY

Dobson Farm "But How Is This Profitable, Anyway?"

The Dobson Farm is a ranch, with rotationally grazed fields for grass-fed beef. The farm uses best management practices and focuses on yearly investment in the land. It is a clear example of brown brain investment.

So, when many years ago, I asked the question, when discussing all the tree planting, "How is this profitable?", Bob took a sip of his coffee and explained:

"There is always water now. The ponds are full, and I can pump from the solar pond up to the central water trough on the big hill. Now, all the animals have access to fresh clean water within a short distance. They gain better that way. It is high and dry up there, and they are out of the mud and their manure is building soil on the uplands. That stuff goes downhill you know? ...and we don't want it in the stream, it'll get into the Ottawa River.

"...It is nice up there for them now. The hawthorns are there, good for them to scratch on. The hawthorns don't mind. They are tough old trees...and good for the loggerhead shrike too; those birds are endangered, you know? They like to impale their prey (frogs and such) on the tree's spiked branches." He laughs.

"And it's protected up there now, the upland shelterbelts were designed to really keep the wind down: the spruce, oak and cedar kind of fit together, you know? It's a jigsaw puzzle of a windbreak.

He motions to the dining room, but I know he means the back of the farm. Also, when the cattle rotate they have water down there too.

Those duck ponds connect the spring with that big reservoir. There used to be just a little trickle there, you know? But we dug it out. Wait a moment!"

I follow him to his office; he rummages around while I scan his yellow-sticky-covered windows. I look closely at one of the faint notes; I could swear it was stuck there in 1982.

"Ah, here we go." He pulls out an old photograph. It shows the stream when it was first dug out. He shows me another that was taken by plane, it was verdant and growing. "See there, that is when I seeded it down in clover and such to stabilize the banks, and we planted a bunch of trees...Put guards on your trees! The mice will eat them up quick; they'll girdle the little trees in the winter.

"Yes, there is a lot of water there now. The trees hold it in and they shade it. Less evaporation and better for the aquatic wildlife, you know? I put that gravity-fed water trough in at the same time. This way when the cattle go down to drink they dip their head in and have fresh running water. Clean water, no

Jig-saw wind break.

Before

After

siltation, the reservoir doesn't have erosion and all those elderberries, cranberries — and grasses too — in the riparian buffer strip, well they keep all the cattle manure from running off into the stream. Yes, that manure is right where we want it: growing the pasture!

"And the cattle don't go lounging about in the mud that used to be down there. Now they are fenced out of the stream and reservoir. This

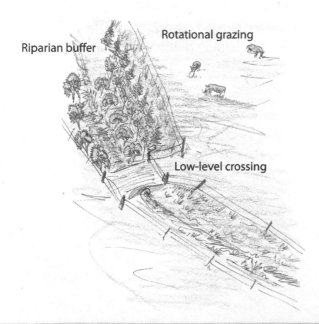

Riparian buffer

Rotational grazing

Low-level crossing

has reduced all the hoof rot and flies. They're much healthier now...cleaner too!" He laughs.

"Seems to be a lot of food for the birds now. Such a beautiful sound in spring."

The glass window ornaments of finches, blue jays and robins bounce rainbows across his face. "We have some of the largest bobolink populations (there is a 3 year bobolink research study underway on the farm now) in the world, they say."

"Well," he said scratching his beard, "I guess there is a lot of potential out there now. Sure, it's been good for the cattle, and we have those elderberries and raspberries and apples we planted. But really, when I think about it, there is, well, just so much potential now. Some of those butternut and walnut and pine are getting big! The farm is pretty well organized around these plantings too. It's much more set up than before. You sure you don't want to be a beef farmer?" I nod knowingly, "There is better water availability, better pasture and better beef than we ever had before. What was the question again?"

He finally took his second cup of coffee.

Guild Enterprise Production

Symbiosis: What Lichen can Tell Us about Farm Profitability

In brief, symbiosis can teach us much about how to organize our garden crops for mutualism and our business enterprises for productivity.

Lichen is actually a symbiotic relationship between an algae and/or cyanobacteria and a fungus. The fungus helps anchor the algae while providing nutrients and water. In exchange, algae provide chemical energy from photosynthesis, and cyanobacteria can fix nitrogen.

Together, these organisms have different forms, functions and biochemistry than individually. This symbiosis gives them a **special niche** in the environment and increases resilience in the face of change. Indeed lichens are primary colonizers in extreme environments, and the map lichen (*Rhizocarpus geographicum*) is the oldest know organism on Earth (about 8,000+ years). Their success is an enviable achievement.

How does this contribute to the profit and success of a farm business? It tells a tale of symbiotic success. Farm systems that maximize symbiosis can improve the farmer's bottom line through increased yield: quality, quantity and and reduced expenses. Expenses are reduced with improved crop functions. Mycorrhizal symbiosis with garden crops is a primary example: crop access to nutrients boosts growth and vigor, and consequentially resistance to disease and pests is improved, helping reduce cost of production.

Symbiotic Farm Enterprises

Let's zoom out and examine how symbiotic relationships can be applied to the farm business as a whole. Although the enterprises of a farm are not organisms, they can be structured to increase beneficial exchanges, including turning waste into fertility, receiving multiple functions from tools and supplies or improving pest management and sharing marketing.

I call this idea guild enterprise production. A farm business composed of three core enterprises that benefit each other by sharing, balancing, inform and cycling resources, information, workload, etc.

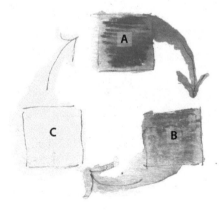

With guild enterpise production a farm is asked to choose three core enterprises and design them to serve each other.

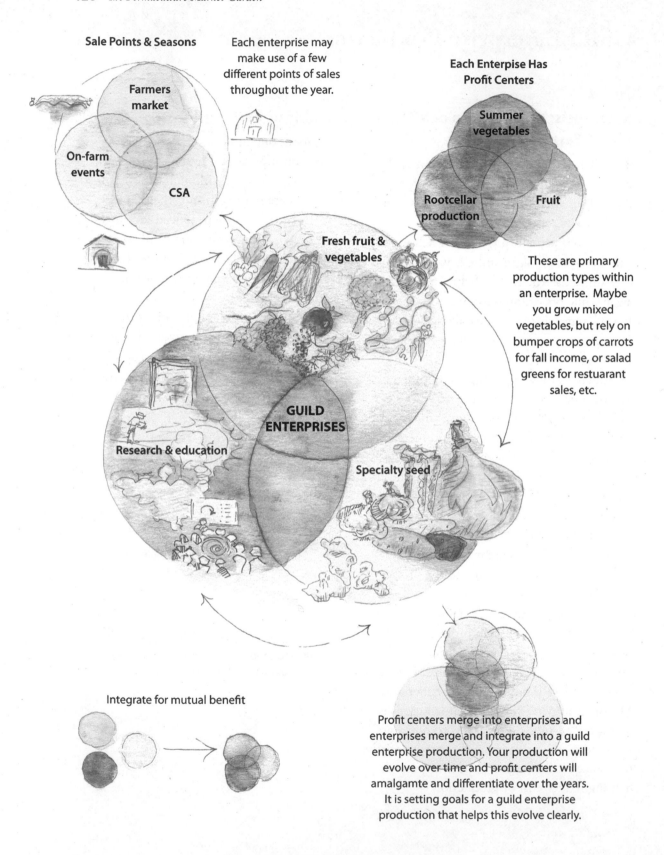

Sale Points & Seasons

Each enterprise may make use of a few different points of sales throughout the year.

Farmers market

On-farm events

CSA

Each Enterprise Has Profit Centers

Summer vegetables

Rootcellar production

Fruit

These are primary production types within an enterprise. Maybe you grow mixed vegetables, but rely on bumper crops of carrots for fall income, or salad greens for restuarant sales, etc.

Fresh fruit & vegetables

GUILD ENTERPRISES

Research & education

Specialty seed

Integrate for mutual benefit

Profit centers merge into enterprises and enterprises merge and integrate into a guild enterprise production. Your production will evolve over time and profit centers will amalgamte and differentiate over the years. It is setting goals for a guild enterprise production that helps this evolve clearly.

Guild Enterprise Production (GEP)

Guild enterprise production builds farm production interrelationships by focusing farm economic production on three core enterprises. Simply committing to three enterpises permits both a limit on exponentially stretching yourself and ensures you have more than one business to help balance and secure your income. This also allows the farm to maximize efficiencies, building beneficial interactions between the enterpises.

Guild Enterpises Interact in a Few Primary Ways

- **Sharing**
  - ○ Improve resource, equipment and tool sharing
  - ○ Share marketing and sales outlets
- **Cycling**
  - ○ Reduce waste by cycling it through other enterprises
  - ○ Better cycling nutrients on the land through diversity
- **Balancing**
  - ○ Balancing our seasonal workloads
  - ○ Balancing product diversity for market resilience
- **Informing**
  - ○ Information gathered in one enterprise helps another

Do You Know? Enterprises have points of sale. These may differ or be similar for your various enterprises. Consider the seasonality of a point of sale to help balance income through the year. Avoid boom-and-bust farming where you only make income from June to November or plan for a different winter income. Also consider how one point of sale may influence other enterprises, such as a CSA program might serve to advertise an off-season workshop series.

Our Guild Enterprise Production

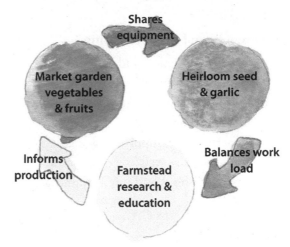

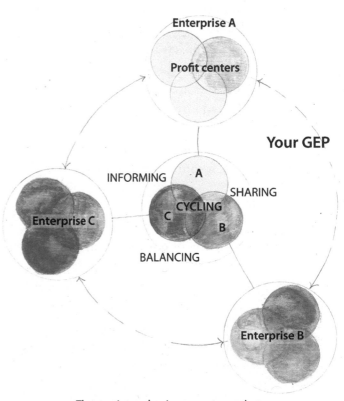

Three primary business ventures that influence each other positively.

Start with one, but consider the others.

Allow them to evolve to be more exactly what is needed.

How We Integrate Enterprises Into a Guild

Equipment and Supply Sharing

• Our seed production and garden business share field equipment and supplies.

• Our infrastructure is multifunctional through time and space. Our barn serves as a wash station, supply storage and garlic curing facility, as well as a space to hold events, workshops and intern trainings.

Enhanced Nutrient Cycling Within the Gardens

• The garlic contributes a significant quantity of mulch into the garden annually, which benefits the fertility management for our organic vegetables.

• Vegetables are very nutrient-demanding, balanced nicely by garlic, which is less nutrient and water hungry.

• Trees and other perennials for fruit and research contribute leaf litter.

Informing

• Both of these enterprises give us the experience for our education and research enterprise.

• At the same time, by investing into research and education, we perpetuate a commitment to improving designs and so end up building efficiencies in our garden and seed operations.

Finding Your Enterprises

Lets go back a few pages. Remember that whole farm mapping and holistic planning? Once your resource base is understood and your holistic goal is made, it is time to look at potential profit centers for your guild enterprise production.

Your GEP is made up of three enterprises (businesses) that can be made up of three productions (aspects of the enterprise). The line between a production and a full blown enterprise can be grey in the early years. Initially your farm will usually be made up of 3 profit centers that eventually become an enterprise as you add other profit centers, which may themselves evolve into an enterprise over the years. Case in point: we used to grow garlic as a market vegetable, and it has evolved into its own enterprise. We also used to grow carrots just for CSA; they now are a cash crop for us in our root cellar profit center. The line between profit center based on an enterprise may also be judged relative to income and the extent to which the production is a unique business. Our garlic production has its own website, whereas our root cellar business is an extension of our CSA season.

Profit centers are microproductions that make up your enterprises. For instance our storage carrots are a profit center within our vegetable enterprise. Profit centers can grow to become more significant within an enterprise and even become the enterprise's primary production. This is how a permaculture market garden evolves. It invests in small profit centers and manages them together,

PLANNING TIP

Our resource base will greatly determine profit centers. It includes the natural, community and personal resources you have available. I wouldn't consider root cellar vegetables as a viable profit center if I didn't have the resource base to support it: good loam soil, a large glacial ridge for my cellar and access to a community demanding winter vegetables.

finding potential budding microenterprises that can help better balance the farm business. Following your holistic goal, observing your land and listening to your community will help you choose which profit centers can expand into enterprises over the years.

How it Starts

I believe every farm should start with three key profit centers. Consider potential enterprises, but leave room to flex your creativity as you gain experience and familiarity with your land, skills and community. These profit centers could be three cash crops: carrots, salad and melons. Or this could be more diverse: apples, garden crops and chickens. This latter is more likely to evolve into three enterprises: orchard production, diversified market garden and pastured meat. The first example should evolve by slowly trialing companionable profit centers until a suitable enterprise emerge. Maybe you begin growing your own vegetable seeds, you see how much you are saving and begin improving your methods until a seed company emerges as an enterprise.

Nurture, Amalgamate or Drop Profit Centers

Nurture *your* best profit centers and allow the others to be amalgamated or dropped. Sometimes the best thing for diversity is simplicity! Make hard choices so your production can improve and not become to hectic.

If you hold on to too many eggs, you will crack them all. Organize them into baskets and then make the basket one of your enterprises.

Do You Know?

Enterprise Self-sufficiency

Enterprises and their profit centers should be self-sufficient. Our seed garlic business holds its own weight. Its income covers its operational expenses, labor and makes profit. This is different than a loss leader, a crop or product within a profit center that may not hold its own but is beneficial to the overall marketing strategy. Peas are needed for CSA, but otherwise they are far down our list of profitable crops.

It is difficult to justify a root cellar or pastured egg business unless they cover costs and make profit. Yet, it is also important to recognize the hidden services these profit centers provide to the overall guild enterprise production. Poultry provides pest control and nutrient boosts, and a root cellar business can help with CSA customer retention and spread out seasonal workload. Despite this, I encourage you to make your profit centers actually cash-in-hand profitable and let these services be cherries on top.

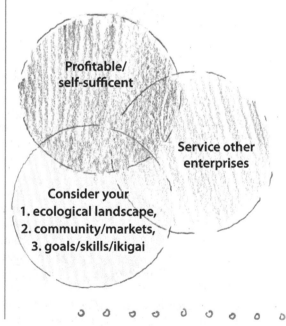

DESIGN TIP

Enterprises must be:

Profitable/ self-suffcent

Service other enterprises

Consider your
1. **ecological landscape,**
2. **community/markets,**
3. **goals/skills/ikigai**

Examples of Guild Enterprises

Pick three: they must be self-suffient, they should support each other … and make sense for you and your ikigai.

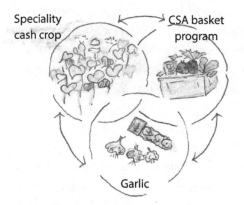

Speciality cash crop

CSA basket program

Garlic

Enterprises that are more alike are easier to manage but offer less resilience. Although different, the production in this example is essentially filling a similar ecological niche.

Winter CSA

Summer farmer's market

Garden-fed eggs

You can diversify your profit by simply extending your season.

This example leaves the realm of just growing plants, however, and adds animals. Now dynamics in the farm ecosystem are increased.

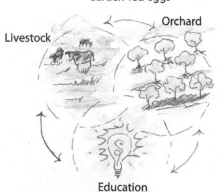

Livestock

Orchard

Education

Lastly, profit can diversify to include animals, annuals and perennials and the exchange of knowledge. Here the potential for service between profit centers is vastly increased.

Guild enterprise diversity increases resilience. Yet, dissimilar productions demand specific knowledge, tools, techniques and labor which can spread farmers thin. We must find enterprises that enhance relationships, build efficiency and savings, and lead to an agro-ecosystem that is regenerative. Don't just diversify without intention and so be spread thin!

Example of Building a Guild Enterprise Production

Imagine a farm that is very remote and heavily covered in hardwoods. The farm couple wants to start a profitable permaculture farm. They could sell firewood. But when we consider the distance to market and the overall low value of firewood, this can be ruled out as a primary profit center.

On the other hand, a business in edible mushrooms is more approachable. Let's break down how this may fit the farm's goals.

The farmers

- The farmer has a background in forestry, has chainsaw skills and enjoys being in the woods.
- They own a car and a small truck, either one can be used for transporting edible mushrooms easily. No investment in a larger vehicle is required.
- They are 3 hours from the city. This high value crop can bring a much greater return per volume than other options, making long-distance travel in a fast and fuel-efficient vehicle likely.
- The land is heavily forested with hardwoods, ideal wood medium for many edible fungi.
- There is demand in the community for fresh mushrooms; no one is meeting this need.

Always consider how your products can be made into a new niche. Assess the demand before heavily investing.

- Cash flow is low and upfront investment to do some initial outdoor shitake trials will be minimal. Financial support through grants provide a means of building infrastructure for year-round production.

Now they must begin to look for other micro-enterprises that can complement this enterprise. Perhaps maple syrup and acorn-fed pigs. Perhaps, clearing a NE hillside for medicinal and edible fruits. They want to find enterprises that can benefit their mushroom production, but also something that works well with the sales model they are developing: Pork, for instance, may have good local demand and also can be processed once a year and delivered efficiently. As for maple syrup, perhaps the local demand is saturated, but a new higher-value product could be developed. Our friends at Wild Branch Botanicals are producing excellent medicinal maple syrups. This is a twist that creates a new niche from a traditional product.

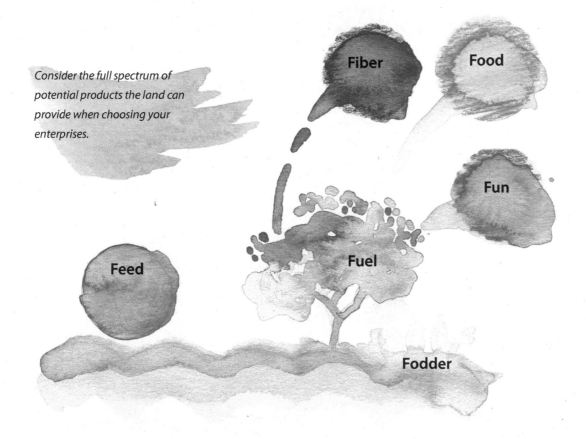

Consider the full spectrum of potential products the land can provide when choosing your enterprises.

CASE STUDY

On-farm Events

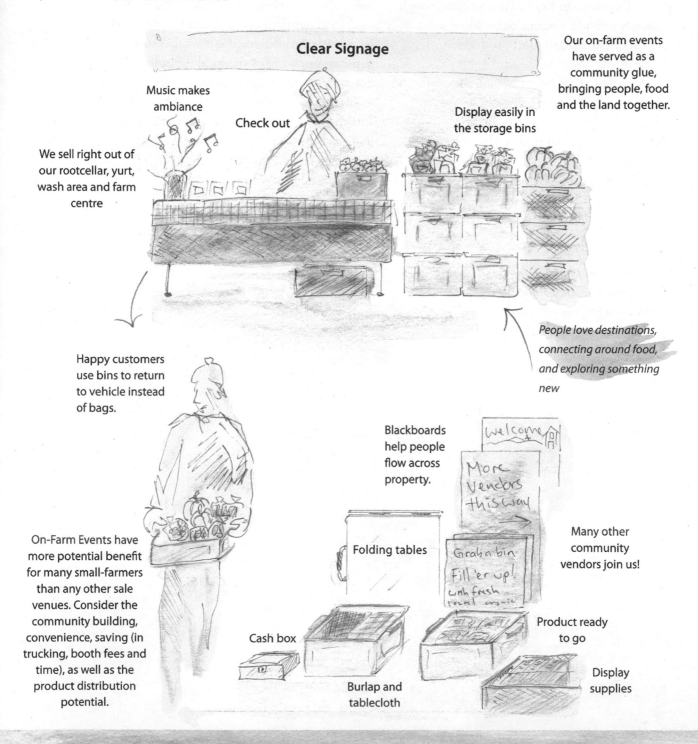

Clear Signage

Music makes ambiance

Check out

We sell right out of our rootcellar, yurt, wash area and farm centre

Display easily in the storage bins

Our on-farm events have served as a community glue, bringing people, food and the land together.

Happy customers use bins to return to vehicle instead of bags.

People love destinations, connecting around food, and exploring something new

On-Farm Events have more potential benefit for many small-farmers than any other sale venues. Consider the community building, convenience, saving (in trucking, booth fees and time), as well as the product distribution potential.

Blackboards help people flow across property.

Welcome

More Vendors this way

Grab a bin. Fill 'er up! with fresh local ensaie

Many other community vendors join us!

Folding tables

Cash box

Product ready to go

Burlap and tablecloth

Display supplies

CASE STUDY: PROFIT CENTER RESILIENCE

Connaught Nursery

Grant looks over the mist as he waters. "The best quality plants I ever produced from these greenhouses..." he smiles at the simplicity, "was when I sleep!"

The Greenhouses:
Here over wintering potted perennials are moved outside and hardy spinach harvested so 20 fruit trees can yield Ottawa Valley-grown peaches weeks before Niagara peaches despite being many zones colder.

Other greenhouses are filled with copious garden starts, early tomatoes, cucumbers, peppers and eggplant.

Connaught Nursery, started in 1982, is a delightful haven where Grant and Dorothy have made a home for themselves and their three daughters. Here a pond drains down amidst vegetables gardens, greenhouses and native plant guilds.

The greenhouses, as the name suggests, were the start of it all, producing garden plants for the local community. The business soon grew into a well-rounded approach to small-scale farming investing in alternative greenhouse production, intensive market gardening and a native plant nursery.

Connaught Nursery is a story of resilience. After years of growing starts for home gardeners, sales were being lost to big box nurseries. In response, Connaught Nursery began emphasizing other enterprises and re-molding old ones. They increased their vegetable garden output, complemented by small fruits and peaches, now reaching peak maturity. Meanwhile, the nursery was refocused on producing quality starts for professional market gardeners and specialty native plants for homeowners.

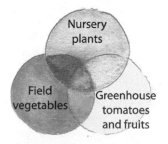

Connaught Nursery's profit guild has evolved over time and now emphasizes professional starts, diversified fruits and vegetables and native plant species. Integrated investment in diverse perennials, flexible infrastructure and tools, along with an efficient management style, has allowed this nursery to easily adapt to change over the years.

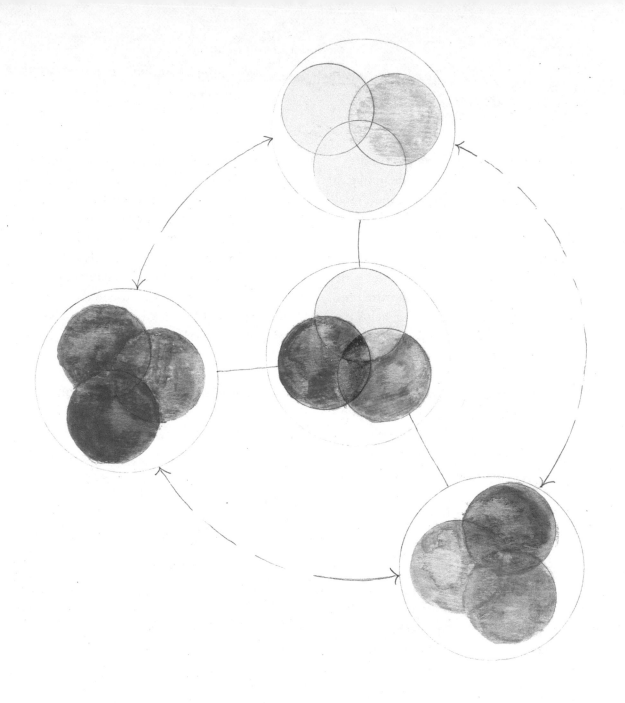

Design Management

Why Design?

Design pulls together all the understanding, mapping and planning. It brings together your goals, business model and available resources. Design should ask the critical questions of what you need from an operation, a tool, a building, in order to draft solutions that do more, better with less. Design should seek to improve time/space/energy efficiency. Design helps avoid costly mistakes by bringing elements together conceptually thus revealing benefits and hindrances prior to implementing them. In this way we can integrate elements to improve benefits and reduce blockages.

Designing for Whole Systems means we are not isolating anything in our farm design process, but rather designing each as part of the whole. Our garden systems are not isolated from the tools we use in them, nor the farm's building, access roads, ourselves, community and land.

DESIGN TIP

Elements of Design

The patterns in nature can inform design. The following elements of design are easily recognizable in nature. They help us conceptualize our goals and visualize design on paper.

- Flow
- Movement
- Nodes
- Opportunity
- Functions
- Size
- Line
- Color
- Shape
- Value
- Space
- Texture
- Hierarchy/leadership and support
- Time
- Scale
- Space
- Contact
- Interaction
- Direction
- Typography
- Dominance/emphasis
- Balance
- Harmony
- Contrast/similarity

Visualize the Elements of Design

Traditional agriculture was labour intensive, industrial agriculture is energy intensive, and permaculture-designed systems are information and design intensive.

— David Holmgren

DESIGN PROCESS: TECHNIQUES AND TOOLS

My Design Tools

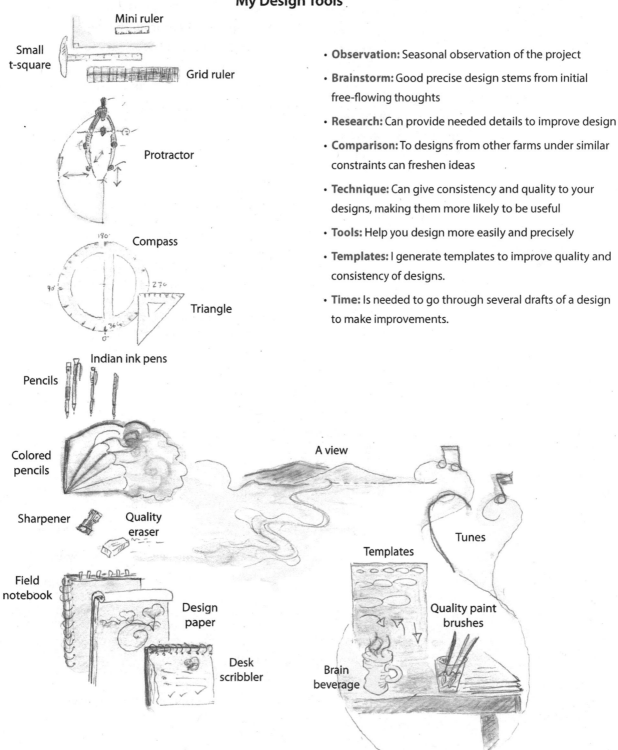

Mini ruler

Small t-square

Grid ruler

Protractor

Compass

Triangle

Indian ink pens

Pencils

Colored pencils

Sharpener

Quality eraser

Field notebook

Design paper

Desk scribbler

A view

Tunes

Templates

Quality paint brushes

Brain beverage

- **Observation:** Seasonal observation of the project
- **Brainstorm:** Good precise design stems from initial free-flowing thoughts
- **Research:** Can provide needed details to improve design
- **Comparison:** To designs from other farms under similar constraints can freshen ideas
- **Technique:** Can give consistency and quality to your designs, making them more likely to be useful
- **Tools:** Help you design more easily and precisely
- **Templates:** I generate templates to improve quality and consistency of designs.
- **Time:** Is needed to go through several drafts of a design to make improvements.

Types of Design

- **Spatial Design:** Relative layout of infrastructure, fields, flows, paths, water movement, irrigation, the big picture's form, function and connectivity.

What is design? It takes many forms, whether in mind, page or laptop, in order to show us how something is made, functions or exists.

- **Specific System Design:** The ice-making system in our root cellar or the seasonal cycle of any crop production system.
- **Crop Guild Design:** The design of a specific grouping of species for common benefit and production: crop analysis, selection, assembly and management.
- **Organization Design:** Field Kits, storage walls, field layout.

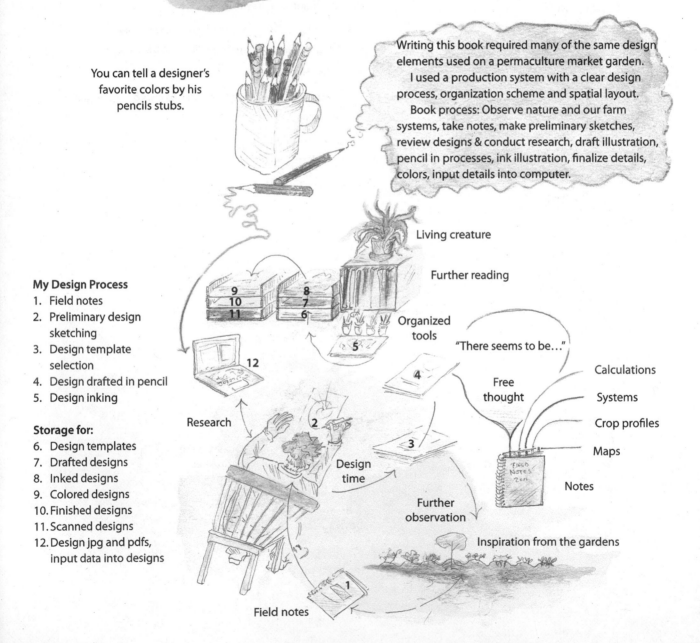

You can tell a designer's favorite colors by his pencils stubs.

Writing this book required many of the same design elements used on a permaculture market garden.
I used a production system with a clear design process, organization scheme and spatial layout.
Book process: Observe nature and our farm systems, take notes, make preliminary sketches, review designs & conduct research, draft illustration, pencil in processes, ink illustration, finalize details, colors, input details into computer.

Living creature

Further reading

Organized tools

My Design Process
1. Field notes
2. Preliminary design sketching
3. Design template selection
4. Design drafted in pencil
5. Design inking

Storage for:
6. Design templates
7. Drafted designs
8. Inked designs
9. Colored designs
10. Finished designs
11. Scanned designs
12. Design jpg and pdfs, input data into designs

Research

Design time

"There seems to be…"

Free thought

Calculations

Systems

Crop profiles

Maps

Notes

Further observation

Inspiration from the gardens

Field notes

- **Management Design:** Designing how something is directed. Record-keeping systems that inform crop planning produce crop schedules.
- → Check out Daniel Brisebois and Fred Theriault's book *Crop Planning for Organic Vegetable Growers*
- **Thought Design:** Designing ways of thinking about farming. For example, creating an "agro-ecological mind set"- considering natural system functions and services when confronted with an obstacle.
- **Design of Design:** We can design for better design. This is part of design management; it is the express management of the design process for consistency, quality and continued evolution for improvement.

Design Process

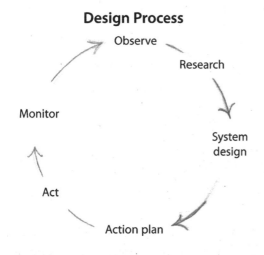

Observe → Research → System design → Action plan → Act → Monitor → Observe

Pit & Mound Index Guild

Living Design & Ecosystem Services

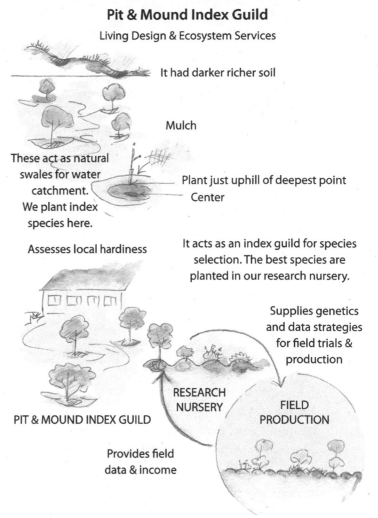

It had darker richer soil

Mulch

These act as natural swales for water catchment. We plant index species here.

Plant just uphill of deepest point Center

Assesses local hardiness

It acts as an index guild for species selection. The best species are planted in our research nursery.

PIT & MOUND INDEX GUILD

RESEARCH NURSERY

Supplies genetics and data strategies for field trials & production

FIELD PRODUCTION

Provides field data & income

We observed the pit and mound terrain: grass grew taller/greener, soil was darker and water collected in the natural depressions. We made use of this by planting fruit trees here. Since this is in zone 1 we developed our Perennial Index Guild here, a planting that helps us test hardiness, grow propagation material and monitor fruit tree cycles of future orchard planted from the chosen successful species.

GREEN THUMB TECHNIQUE

Asking "Why?" 7 times

Asking "Why?" can reveal design solutions. For example: "I want to design a better barn space" "Why?" "Because, I want make it easy to find tools." "Why?" "Because we have many tools that are needed." "Why?" "Because every day we do many different jobs. "Why? "Why?" "Why?" This resulted in designing field tool kits that assemble needed tools and supplies for routine jobs. "Why?"

Principles for Design Management

We must integrate design management into the whole farm. The following are principles, grown from permaculture and our own design management process, for the whole-farm design management.

Rule of Three

Douglas Adams says the meaning of life is 42, I say it's 3

- Three is the first number where there can be complex relationships. It is a balanced unifying number in all our design (DMZs, TRIADS, CROP GUILDS, see Glossary).
- When you give yourself design constraints, it simplifies the design process. The rule of three has greatly enhanced our design management of the farm.
- The number three is quite popular in society and nature. The formula for a good story: beginning, middle and end; the three primary colors: red, blue and yellow; and the natural states of water: liquid, solid and vapor.
- We also use multiples of three — 6, 9, 12 — in our designs.

Use the Full Color Spectrum

- We use the full spectrum of colors to our advantage. For example, an array of colored duct tape helps us differentiate between the field kits we have assembled for all routine jobs. We also use the colored tape for To Do stakes that help us prioritize irrigation movement (blue), seeding and planting (green) and bed preparation (red).

Redundant and Regenerative

- **Redundant:** Design for redundancy to minimize system failure. We have many ways of maintaining important crop needs: soil moisture is obtained through rain, pond pump and gravity irrigation, but we also mulch, improve soil organic matter, and reduce wind evaporation with cover crop intercropping. It's NASA for Farmers.
- **Regenerative:** A healthy soil with balanced organic matter, minerals, air and water pores, biological life and aggregated structure can cycle nutrients quickly for improved production after a disturbance like drought or tillage.

The more important the function, the more redundancy is imperative.

Aesthetic and Useful

- **Aesthetic:** If a design isn't appealing to the eye it won't be respected over time. Structures, gardens and spaces (no matter the budget) can be designed to be appealing and so will be respected by those who pass through a farm and those who become the successors to your efforts.
- **Useful:** A hedgerow between fields may be removed by future farmers. But if it has high-value timber and tasty pears or is a well-tended sugar maple lane for syrup, it will pass the test of time.

Integration

- When we integrate farm production, it means there is potential symbiosis created between cash crops, cover crops, animals and trees for improved agro-ecology.

  REMEMBER: Enterprises can be integrated to share resources, cycle outputs/inputs and inform decisions. We call this guild enterprise production. DMZs are a way of organizing space for integration across the farm.

Integrating Flows Between DMZs

- We should design to integrate flows between DMZ.
- Inputs, output and waste integration: waste becomes an input to increase output.
- For instance, our wash station becomes a collection point for water because it is integrated with a pond to catch and use the water.

Whole Farm Design Management

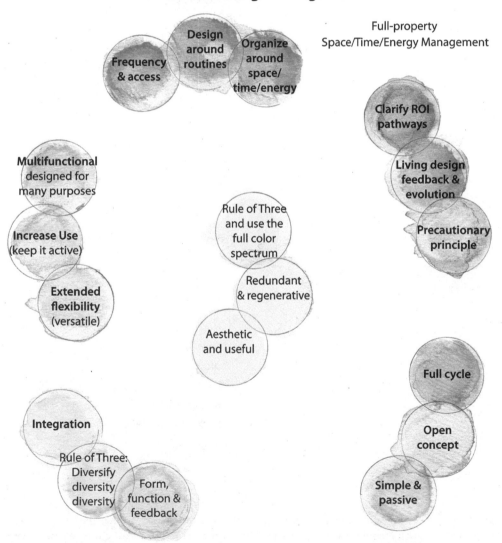

Full-property
Space/Time/Energy Management

Frequency & access

Design around routines

Organize around space/time/energy

Clarify ROI pathways

Living design feedback & evolution

Precautionary principle

Multifunctional designed for many purposes

Increase Use (keep it active)

Extended flexibility (versatile)

Rule of Three and use the full color spectrum

Redundant & regenerative

Aesthetic and useful

Full cycle

Open concept

Simple & passive

Integration

Rule of Three: Diversify diversity diversity

Form, function & feedback

Diversity, Diversity, Diversity

- Diversity is more than the different crops you grow. Consider the diversity of garlic types, sizes, and the marketing styles that can be used at your market booth.

Diversity of labels

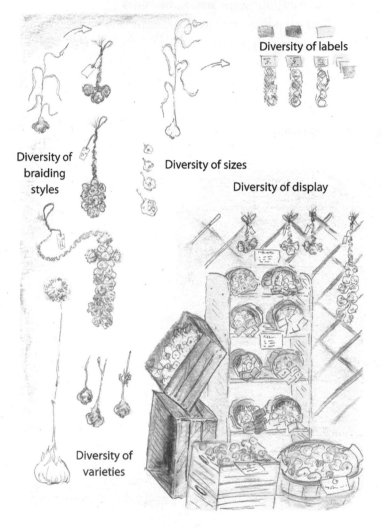

Diversity of braiding styles

Diversity of sizes

Diversity of display

Diversity of varieties

Form and Function

- Everything has form and function. When we design with these in mind we can create beneficial interactions by balancing for the best interrelationships. For instance, taprooted oaks work well with wide-rooted apples, and both are enjoyed by pigs.

DESIGN TIP

Make a design notebook, divide it by colors and for purposes (sketches, maps, crop notes) to help you brainstorm, observe and design.

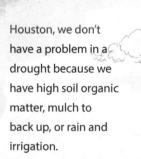

Houston, we don't have a problem in a drought because we have high soil organic matter, mulch to back up, or rain and irrigation.

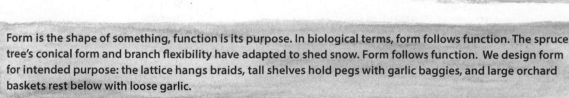

Form is the shape of something, function is its purpose. In biological terms, form follows function. The spruce tree's conical form and branch flexibility have adapted to shed snow. Form follows function. We design form for intended purpose: the lattice hangs braids, tall shelves hold pegs with garlic baggies, and large orchard baskets rest below with loose garlic.

Multifunctional

Design for multiple functions. Cover crops improve the soil, and intercropped in alternating beds with summer salads, they also serve to cool the greens and reduce heat stress.

Do You Know?

Our rubber bins on the farm are a great example of a multifunctional tool. We happily invest in them because they are so universally useful.

Multiple Functions

We use bins for many purposes and re-purpose them based on wear and tear.

Old bins are used for harvest

New bins go to market

Bins with worn-out handles store supplies in barn

Easy way to clean seed BIN + BIN + FAN

They also become the second tier for market display

Perfect for fast display of products at on-farm markets

Bins serve as ice forms in winter to make 1,000 cubic feet of giant ice cubes that are stored to cool our cold storage in summer.

Increase Use

- Plan for a tool or service to be in more constant use. Many tools or infrastructure remain unused for months at a time, like summer production buildings. Our wash station serves as a space for transplants in spring, vegetable processing in summer and ice-making in winter.
- Some supplies may be unused for periods of time. We move weed barriers and row

What to Design for...

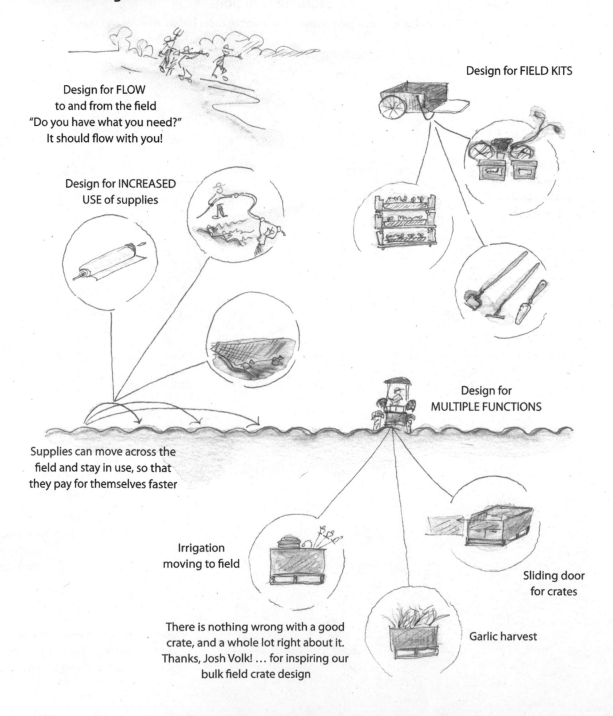

Design for FLOW
to and from the field
"Do you have what you need?"
It should flow with you!

Design for INCREASED
USE of supplies

Design for FIELD KITS

Design for
MULTIPLE FUNCTIONS

Supplies can move across the
field and stay in use, so that
they pay for themselves faster

Irrigation
moving to field

Sliding door
for crates

Garlic harvest

There is nothing wrong with a good
crate, and a whole lot right about it.
Thanks, Josh Volk! ... for inspiring our
bulk field crate design

cover across the field over the season, never rolling it up and storing it. If you are rolling row cover up in the middle of summer, you haven't maximized its seasonal use.

Extended Flexibility

Design for unknown uses and future possiblities.

- Make the design fit the space.
  - We designed our root cellar packing table so it could be turned in our packing bays. This way it could be used up against a wall (its intended placement) or turned to run parallel to our main runway. Many years later we found this was a good idea because now we turn it for on-farm community events and use it as a sales counter.

Bagging Station

- Keep the design general but useful.
  - Our packing table has many special features; there are boards that can be removed so we can conveniently drop spoiled or C grade root vegetables into crates below. However, it would be easy to cut and screw two 2×6 boards to fill in those spaces. This set up is extremely specialized, allowing us to do a very specific sorting job and yet it remains very simple to return the table to a general state and be repurposed.

Frequency and Access

- Design for improved access. For instance, low-growing or early harvested crops beside larger perennial crops to improve access for perennial harvest. Early crop beds are already harvested and become accessible beds later.
- Design around frequency. Permaculture zones for the property and garden are good examples. We should assign productions or projects that require frequent visitation in zones where personnel already visit frequently.

Lay out your farm for frequency and accessibility

Permaculture zones divide up property space according to how often different areas are frequented, and operations are assigned to zones based on how often attention is needed (daily, weekly, monthly, etc).

Kit Organization for Efficient Management

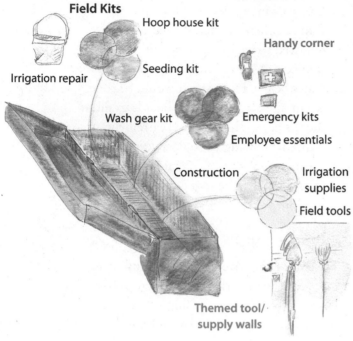

Field Kits

- Irrigation repair
- Hoop house kit
- Seeding kit
- Handy corner
- Wash gear kit
- Emergency kits
- Employee essentials
- Construction
- Irrigation supplies
- Field tools
- Themed tool/ supply walls

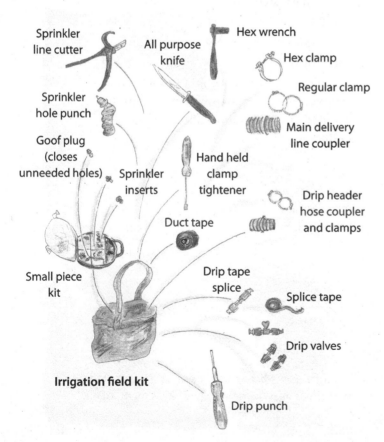

- Sprinkler line cutter
- All purpose knife
- Hex wrench
- Hex clamp
- Regular clamp
- Sprinkler hole punch
- Main delivery line coupler
- Goof plug (closes unneeded holes)
- Sprinkler inserts
- Hand held clamp tightener
- Duct tape
- Drip header hose coupler and clamps
- Small piece kit
- Drip tape splice
- Splice tape
- Drip valves
- **Irrigation field kit**
- Drip punch

Design around Routines

- Most market garden operations are quite routine. Once you have been producing for three years, you can identify what you do, when you do it and how it is done. As such it becomes a matter of designing around these routines. The following are designed from routine operations:
  - Operational protocols: a sheet that says exactly how it is done
  - Tool kits: a bag with all the needed tools
  - Seasonal schedule: A plan of when everything is done
- Protocols, tool kits and seasonal schedules ensure operations are assembled for efficiency.

Organize around Space/ Time/Energy

These are interconnected in everything.

- **Space:** Farms can be big, medium or small. It is how we use these spaces that makes the difference. Turn space into place so you can better familiarize yourself with its potential and use it more efficiently.
- Our permabed system organizes land for better production by dividing space and better understanding the ecological character.
- **Time:** It is of the essence on all farms. Design for improved use of time. For instance, we alternate the maturity of crops so as to more easily and rhythmically cover crop our fields. Fast-maturing crops are seeded to summer cover crops and serve as alleys to facilitate harvest of slower maturing crops.
- **Energy:** Our field kits save us time and energy as we move through space with pre-assembly of needed tools for specific jobs. Don't waste calories to run back for a

self-tapping screw when repairing your high tunnel.

Clarify ROI Pathways

- It is important to clarify how a project or process returns on its investment (ROI).
- Obviously the sooner it does so, the happier we will be.
- However, we have many ongoing projects that return on a longer-term and in less visible ways. By clarifying their ROI pathways, we can set doubts aside and continue to invest in them.
- For instance, I know planting cedars along my driveway to fill in the gaps in our hedgerow and reduce wind and snowdrifts is a good investment.
- We also clarify that equipment for conservation tillage is improving our yields through soil life conservation. Ex: A power harrow doesn't create the plow pan a rototiller will.

By clarifying ROI for less obvious investments, we make better decisions for a sustainable operation.

Living Design Feedback and Evolution

- Designs must be perfected through use. Until a model design is put into the context of its intended purpose, it remains untried and so cannot be fully designed for maximum system efficiency.
- Observe for feedback from the living design and redesign for improvement.
- Production improves through feedback loops that point out obstacles and successes. Walk your land regularly to watch for system feedback. Anything you do in the garden will provide feedback. How well did it germinate? Why? Is this process working well? Why not?

Precautionary Principle

- Exercise precaution in design by trialing ideas first in controlled areas before they are used more fully.
- We trial new varieties of perennial ground covers in research gardens where we easily observe and control them.

Do You Know?

Higher perspective helps find solutions
A different view gives new perspective. Down amongst the beds, it is easiest to learn about crops, weeds, pest, harvest readiness and study bed-scale solutions. However, high up on hillside, a barn loft or rooftop you can better see relationships between your garden and your property, between your property and the surrounding countryside. Here you can better judge the crops rotation in your land and fine-tune landscape-scale solutions.

Design for full cycle

Full Cycle

- Design with the full cycle in mind to improve management of future stages in advance.

- Understand the crop's full production cycle and also its life cycle when designing for potential crop services.

→ See guild crop rotation later.

Seed to Salad

Understand the cycle, the order of operations, the stages of production

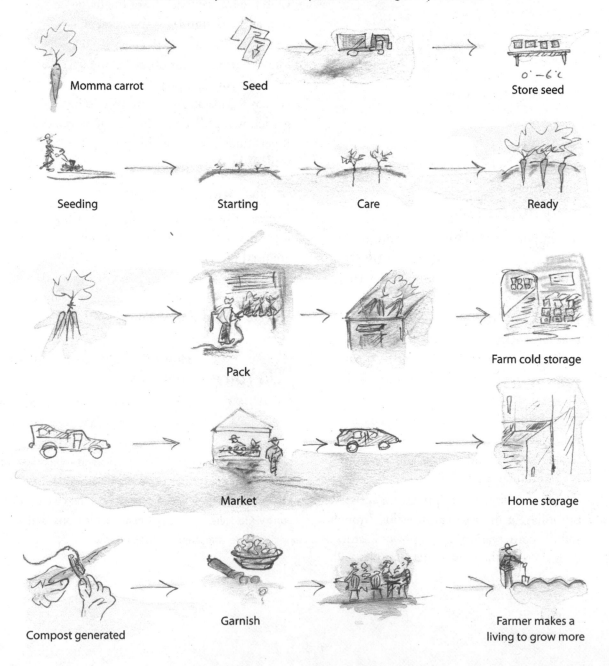

Momma carrot

Seed

Store seed

0° – 6°c

Seeding

Starting

Care

Ready

Pack

Farm cold storage

Market

Home storage

Compost generated

Garnish

Farmer makes a living to grow more

Garlic Crop Production Cycle

Design systems to be full cycle

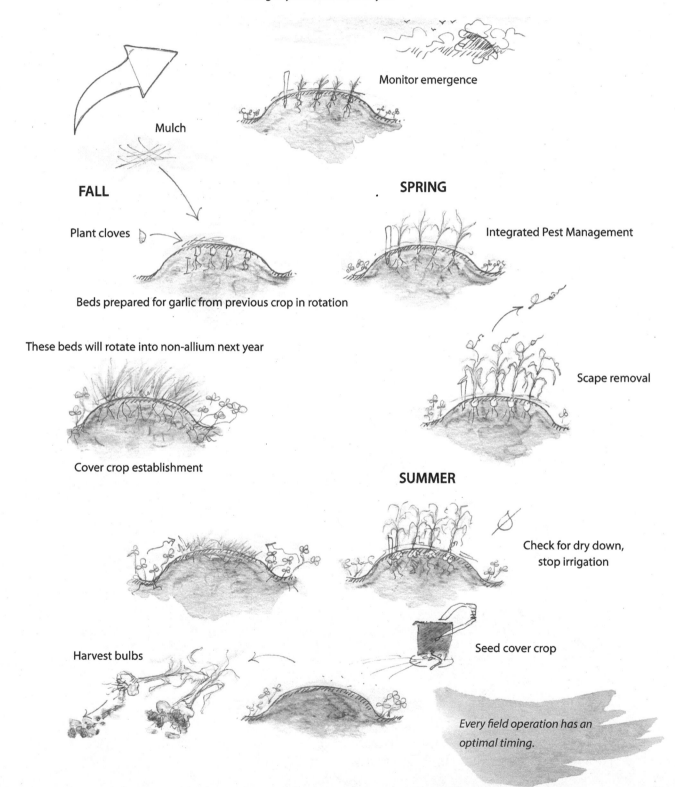

Monitor emergence

Mulch

FALL

SPRING

Plant cloves

Integrated Pest Management

Beds prepared for garlic from previous crop in rotation

These beds will rotate into non-allium next year

Scape removal

Cover crop establishment

SUMMER

Check for dry down, stop irrigation

Harvest bulbs

Seed cover crop

Every field operation has an optimal timing.

Full-cycle, Multifunctional & Integrated Design

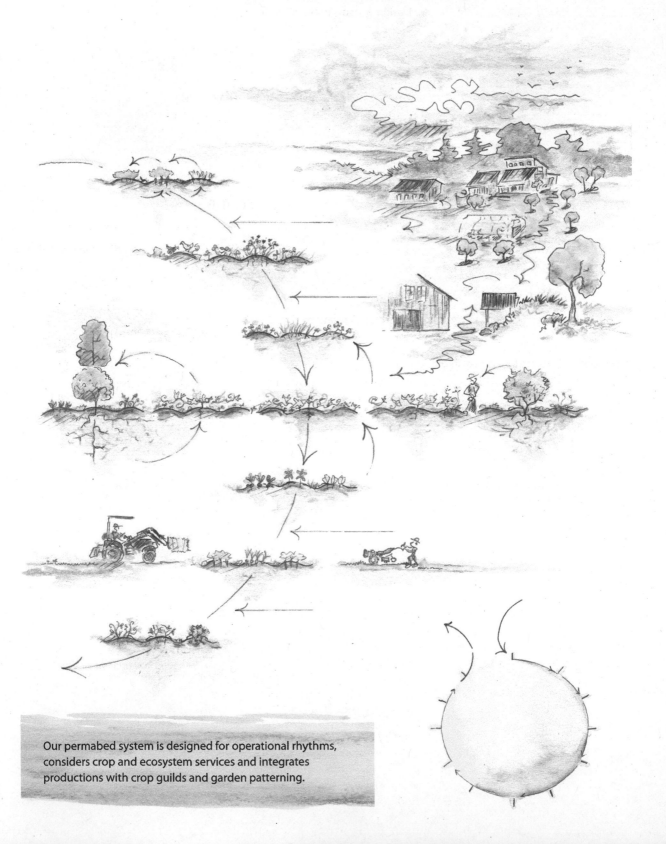

Our permabed system is designed for operational rhythms, considers crop and ecosystem services and integrates productions with crop guilds and garden patterning.

Open Concept and Flow

- We designed our barn with open concept; you can drive around it, right through it and move easily within its open bays. This building could be used for many purposes. The open concept and flow increases its usability and gives it extended flexibility.

Root Cellar & Barn Design Management Zone

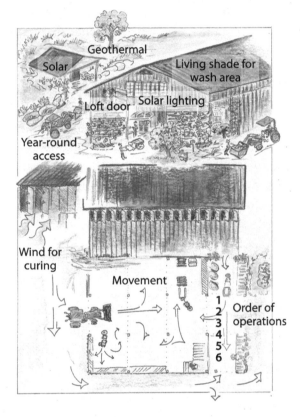

Laneway Farming

Simple and Passive

Simple is best. Simple and passive systems don't fail and cost less to operate. Our root cellar is kept cold using ice from the winter which we store at the back. This ice melts causing cooling. If the weather gets hotter, the system adjusts automatically, increasingly cooling as the ice melts in response to warmer weather.

Design management is better for the farmer's wallet and your farm's ecological footprint.

CASE STUDY

Our Passive Root Cellar/Cold Storage

Passive Cold Storage

Our root cellar is a year-round passive cold storage. It is cooled by ice which we generate naturally during the winter months. We make 1,000 cubic feet for an annual cost of $500. It maintains our desired temperature and humidity conditions all year.

Multiple Functions

1) Stores summer vegetables, 2) becomes a winter root cellar, 3) provides geothermal temperature moderation, 4) stores and uses ice cooling, 5) thermally regulates itself 365 days a year, 6) serves as an on-farm store, 7) is a nursery tree storehouse in spring, 8) is a research project, 9) water collection point for our soaker-pond research nursery, 10) underground dance hall, 11) hobbit hole, and 12) apocalypse shelter.

Design

When we began planning our root cellar, we placed various constraints on its design. We wanted to avoid damaging the century-old maples on the ridge into which it is dug. We wanted a space that was large enough for future yields of root vegetables through the winter and that could be cooled using

> We need quality design management to help build local solutions from a global knowledge base. Model designs exist all over the world, ready to be adopted by you and adapted to your situation.

alternative technologies because we are off-grid.

Thanks to Chris Chiasson at Whole Farm Services, countless brainstorm sessions and much trial and error we began a journey that would epitomize design management for us. Here are some of the ways our root cellar has done this.

Cycling Energy and Matter

Production processes have Inputs, Outputs and Waste. The projects and operations within a DMZ work to integrate these. The waste of one process becomes an input for another. The output of one process is benefiting from another's design.

- We solar-pump water to make the ice that cools our cellar all summer
- Our root cellar melt water feeds our soaker-pond nursery
- We mulch the nursery with small square bales used inside the cellar to insulate the top of the ice mass when the ice is melted in October.

Leave Cycles Open for Flexibility

The melt water from our root cellar's ice chamber, the rain catchment from our barn roof and the wash water from the wash station under the barn are all collected and run to a catchment pond, which slowly feeds the moisture to our soaker-pond nursery. When we built our cellar and barn we didn't know where the water was going to be used. We understood that it should be captured, stored and eventually connected. When the time came, we began to develop our soaker-pond that feeds our nursery. A little design forethought meant we were able to connect and use this resource later. Leave cycles open for future uses through flexible design.

Managing the Design

It took many years of trialing, observing and reworking to come up with a design that allows us to cool our cellar passively and affordably. Here is a breakdown of our ice-making system.

Ice-making System

1. Wash tables become ice shelves
2. First-in carrot storage bay becomes ice chamber
3. Ice filling accordion (drip irrigation) is easy to move
4. Drip filling is a breeze
5. Pressure regulator optimizes bin filling
6. -15 to -35°C is best
7. Fill to 3" below lip
8. Roll 'em in!
9. Move 'em on!
10. Pop 'n flip!
11. Stack 'em up! Don't make them heavier than you like to lift
12. Start at one end and build consecutive wall to the top
13. Step up when needed
14. Anti-slip mat
15. Layer snow as mortar
16. Clean up
17. Small squares bales seal the entrance to ice chamber
18. Tarp cover
19. Reducing heat
20. Melting ice
21. Collecting water
22. Research nursery pond
23. Veggies in
24. Small square bales mulch the roof
25. Living insulation
26. Harvest crates used to elevate ice
27. Rubbermaid bins become ice buckets
28. Frost-free hydrant can be used at -35°C

Design Management

Design management is the emphasis on design for better management. It is the management of the design process, the overarching goal of improved times/pace/energy productivity, and a commitment to ever-better design by allowing design the opportunity to evolve in an organic way.

Let's look at these individually:

Design for Better Management

We can greatly improve our management of the farm with designs. It is easier to manage my wash station when it has a well-thought-out layout and the wash tables, bin storage

Design Management

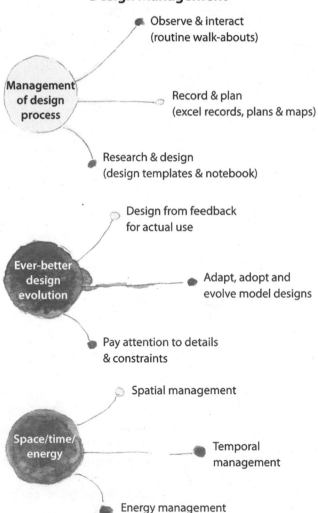

Management of design process
- Observe & interact (routine walk-abouts)
- Record & plan (excel records, plans & maps)
- Research & design (design templates & notebook)

Ever-better design evolution
- Design from feedback for actual use
- Adapt, adopt and evolve model designs
- Pay attention to details & constraints

Space/time/energy
- Spatial management
- Temporal management
- Energy management

Wash Station Design
(Design for space/time/energy)

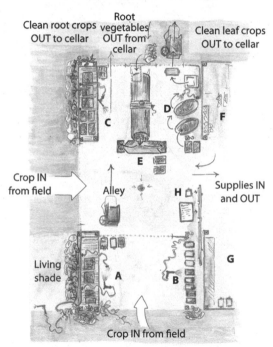

Clean root crops OUT to cellar
Root vegetables OUT from cellar
Clean leaf crops OUT to cellar
Crop IN from field
Alley
Supplies IN and OUT
Living shade
Crop IN from field

Design of our wash station for smooth crop flow, versatility of infrastructure. Overhead hose attachments make it easier to manage busy harvest days.

A) Primary wash table
B) Bin & crate storage
C) Secondary wash table
D) Washtubs
E) Harvest headquarters
F) Employee shelf and wash gear
G) Post-harvest supply storage

and harvest lists work together as a complete system. Obviously farms are doing this all the time. The question here is why aren't we doing it more intentionally as a matter of course.

Design for Organization

Field kits and supply centers save us time and energy because we always have what we need on hand. They make it easy to fill extra time with odd jobs and prevent runs back and forth to the barn.

Field kits are kept on their own shelf with a check-in/out table

Field kits have all the essentials for doing a routine job. The bag and its contents are color coded. If you need a tool that isn't in the kit, get one specifically for the kit and color code it with duct tape.

In the field we organize our supplies using wooden crates placed in front of our perennial triads. We call these supply centres. Tool kits can be placed here while we work, but they return at day end. Supplies: row cover, drip irrigation, sprinkler stakes, wire hoops, etc. remain in supply crates for easy access.

Supply crates return to the barn when a seasonal routine is finished.

Smaller farm solution? Consider a supply cart, that your kits easily fit into.

1. Management of Design Process

Design is far too important to just do willy-nilly. We must have a well-organized approach to designing to help us do it in a uniform and timely manner.

It is good to have systems that integrate observation, record keeping, research and actual hands-on design. We employ the following:

- Routine walkabouts
- Templates

- Design notebook
- Excel record sheets and crop plans
- Field maps
- A design notebook is essential for free-flow thought, observation and research and then concrete design ideas.

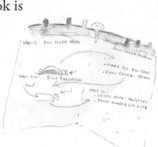

Routine Walkabout

Observe, Record, Prioritize

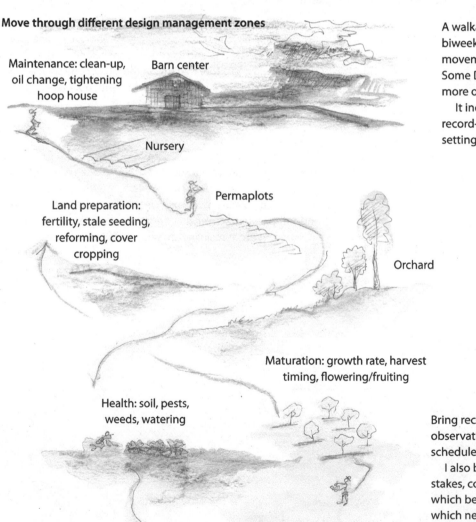

Move through different design management zones

Maintenance: clean-up, oil change, tightening hoop house

Barn center

Nursery

Land preparation: fertility, stale seeding, reforming, cover cropping

Permaplots

Orchard

Maturation: growth rate, harvest timing, flowering/fruiting

Health: soil, pests, weeds, watering

A walkabout is a weekly, biweekly and monthly movement around the farm. Some DMZs should have it more often.

It includes: observation, record-keeping and priority setting.

Bring record kit: record books, observation/record maps, to-do schedules.

I also bring field operation stakes, color-coded to prioritize which beds need to be reformed, which need irrigation and which need weeding.

Example: Design Templates for Observation and Design

- Any design work I do on a regular basis deserves its own template to help the process and make it more consistent. See, for instance, this template to help me record and design layouts for garden plots.
- Excel spreadsheets are useful for records.
- I also use Excel for mapping gardens and properties.
- Our system of permanent raised beds make it easy to allocate space for crop trials.

Permaplot Template

Design Template Used to Brainstorm High Tunnel Rotations

A system must be designed for four seasons.

This includes thinking through what needs to be in place for a desired production at a desired time.

In full-cycle design, we discover opportunities for improved management and savings from space/time/energy relationships.

For instance, we used to try and cram our hoop house with crops in every season and then fill it with compost and quickly turn it over for more production. We save a lot of time by moving our hoop houses to new fertile ground and cover cropping the previous ground for slow fertility return.

Do You Know?

Understand your plants' vulnerability.

Young germinants are extremely fragile in the wind-whipped, sun-scorched, rain-spattered environment of a typical garden.

2. Commitment to Ever-better Design Evolution

Design for feedback. A design should not be a final product. Our wash station, for instance, should be ready to receive the feedback of *its actual use* and be remodeled to suit the processes better. Consider how to make your design flexible for improvement.

People satisfaction, operation efficiency and Soil improvement are benchmarks of a better design

Ever-better Design

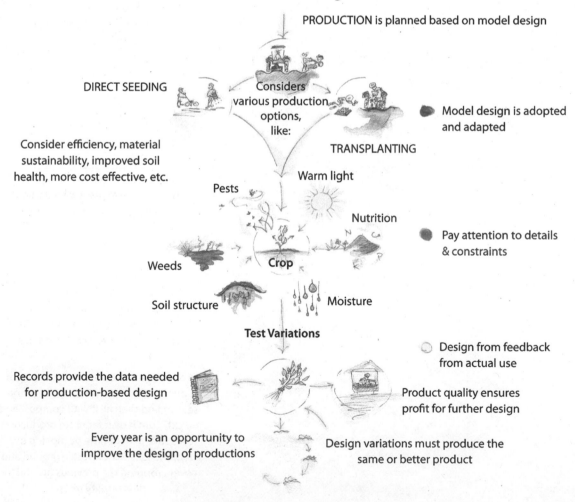

PRODUCTION is planned based on model design

DIRECT SEEDING

Considers various production options, like:

Model design is adopted and adapted

Consider efficiency, material sustainability, improved soil health, more cost effective, etc.

TRANSPLANTING

Warm light

Pests

Nutrition

Pay attention to details & constraints

Weeds

Crop

Soil structure

Moisture

Test Variations

Design from feedback from actual use

Records provide the data needed for production-based design

Product quality ensures profit for further design

Every year is an opportunity to improve the design of productions

Design variations must produce the same or better product

DESIGN TIP

Design work requires time: set aside 1 hour/week; space: sit down somewhere enjoyable; and energy: bring your brain fired up and the tools of the trade

How can farmers better use space, time and energy now and into the future.

Model design adopt and adapt: Model designs are those we adopt from other farms because they seem like a good idea. Model designs must be adapted to your goals, land and situation. A design innovation from one farm will never serve you in the same way. We all have different microclimates, ecologies, tools, customers, etc. It is our role as design managers to take these model designs and work with them, trialing, observing.

Pay attention to the details: Better design is in the details. What makes a system run smooth is that packet of shearbolts on the tractor ready for when one breaks, clear labels on supply bins, easy communication for jobs to be done. For instance, we use to-do stakes to help delineate routine tasks in the field. Stakes with red tape mean the bed should be reformed,

PLANNING TIP

Pay Attention to Constraints

When designing a crop's production system, pay specific attention to the natural constraints: preferred pH, soil moisture regime, maturity, etc. We must also consider other constraints: acreage availability, equipment costs, skills needed, etc. See if you can find key constraints (weak links) and design solutions.

Principle   ### Design With the Permaculture Principles in Mind

Catch & Store Energy

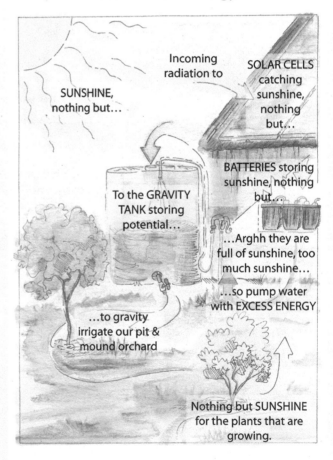

SUNSHINE, nothing but…

Incoming radiation to

SOLAR CELLS catching sunshine, nothing but…

To the GRAVITY TANK storing potential…

BATTERIES storing sunshine, nothing but…

…Arghh they are full of sunshine, too much sunshine…

…so pump water with EXCESS ENERGY

…to gravity irrigate our pit & mound orchard

Nothing but SUNSHINE for the plants that are growing.

Design Management

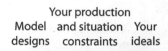

Model designs   Your production and situation constraints   Your ideals

Designable: Crop production cycle, a barn layout, or even a new workbench

NEW DESIGN
Potentially better design

A MODEL DESIGN

Any design you begin to adapt for your projects & situations, gleaned from other farms or inspired from other domains. It may have been used and trialed before and is now being applied in your context to meet your needs.

Design management takes model designs, your constraints and your ideals, and applies them to any designable (say crop production cycle) to produce a better design.

Observe, Record and Research (Design Notebook)

New DESIGN

Production using design

VARIATION of NEW DESIGN
ready for trialling

How well does it reflect design principles?

How well does it meet needs?

Cost/benefit analysis?

How well does it integrate with other projects and systems within the DMZ?

ASSESS USES, EXHANGES, RECYCLING

DMZ RELATIONSHIPS

Remodeled design

Outputs to other DMZ or off-farm

DESIGN TIP

Always consider how a NEW DESIGN variation will interact with other elements within a DMZ. What new waste does it produce? How can it use other project outputs? Which resources and tools can be shared? DMZs are about mutualism, and changing one aspect requires consideration of its effect on others.

3. Improved Time/Space/Energy Productivity

Everything we do on the farm is affected by time, space and energy. There is a distinct savings in labor, money and resources when any of these elements balance more favorably. For instance, do a job in less time (economy of motion) or do it with less space (integrated production).

Time management: Consider how many distinct motions are needed to transplant an onion.

Energy management: Consider inputs needed to plant this crop.

We must consider how much space is needed in time and with energy. For instance we leave more space in our row crops (potatoes, peas, beans), but it saves us energy [labor] and gives a great opportunity for early [time] cover crop establishment between the rows. We come out ahead with this design. Because three different crops can be designed for similar field management (such as hilling), we can plant more diversely without loosing efficiency. This is called umbrella management.

Umbrella Management

Three crops grown in triads with dissimilar management can form a triad with similar management.

Space management: Consider how much space is allocated to each plant and how we can better use that space.

Quality management: As market gardeners we must be interested in quality. Consider

a spring turnip. It is a unit of energy. If it is half-riddled with maggots, then this lower-quality turnip is undesirable. It therefore justifies the increased energy of laying row cover to ensure it fetches the correct price.

Money management: We discuss this elsewhere. But here at least we can say that we must manage design so that is it financially viable. A design that uses little energy, space, time and produces a quality crop is no good if there is no demand for it.

Design management zones: Consider how best to use space relative to time and energy. One way to manage space on a property scale is through the designation of these design management zones, distinct places across your property with specific production centers. Consider our farm center DMZ with its structure, productions and process.

Seasonal schedules dictate what activities to expect in any one season. This helps us to look for opportunities to balance our agenda through the year, so we don't have a spring bottleneck or summer frenzy or jobs left undone in fall.

By defining exactly what goes on within a DMZ — the tools, skills and timing of operations — we can be better prepared for efficiency by design managing this piece of our operation. For instance, we are working to create field kits, protocol sheets and seasonal schedules for all our DMZs.

This help us reduce waste from running around the farm without clear job understanding, tools and timings. When moving around the farm, we do so intentionally and with what skills and tools we need.

Make connections between the spaces of your farm. Make them cooperative places.

Farm Centre Design Zone

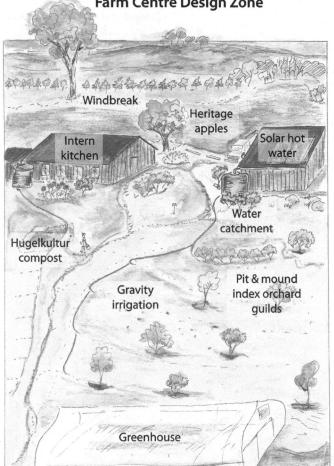

Windbreak

Heritage apples

Intern kitchen

Solar hot water

This area is the most trafficked zone. It is the connective tissue between living and working.

Water catchment

Index guilds can be planted in hugelkultur compost.

Hugelkultur compost

Gravity irrigation

Pit & mound index orchard guilds

Greenhouse

A model design is any idea from another farm (or elsewhere) that could be adopted and adapted to your situation. This could be an entire salad production system or just a salad cutting tool, a marketing strategy or an organizational concept like DMZ, it could be the plans for a curing shed or a cover cropping strategy.

Design Management Zones Keep Time/Space/Energy Flowing

DMZ should inform each other's design, use each other's waste and share resources

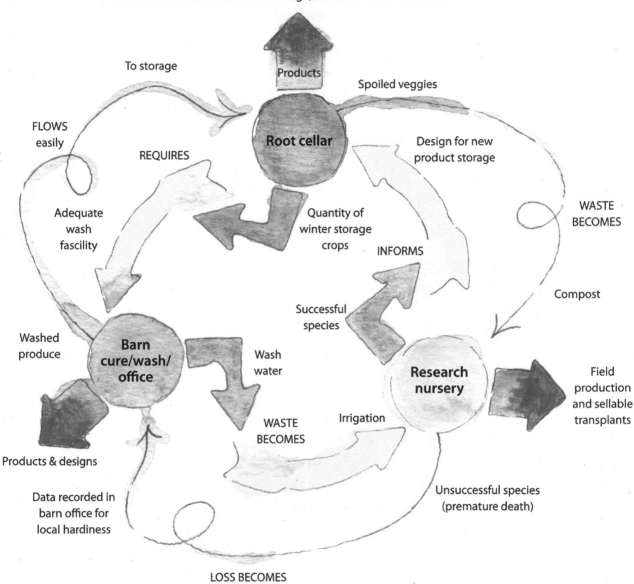

Products

To storage

Spoiled veggies

FLOWS easily

REQUIRES

Root cellar

Design for new product storage

WASTE BECOMES

Adequate wash fascility

Quantity of winter storage crops

INFORMS

Compost

Washed produce

Barn cure/wash/ office

Successful species

Research nursery

Field production and sellable transplants

Wash water

WASTE BECOMES

Irrigation

Products & designs

Data recorded in barn office for local hardiness

Unsuccessful species (premature death)

LOSS BECOMES

Principle   Design considers the life cycle of the organism (crop, animal, tree). You cannot improve design unless you know the natural rhythm of a crop and strive to work within it.

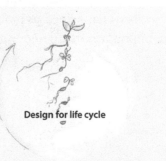

Design for life cycle

12 Permaculture Principles
* From David Holmgren

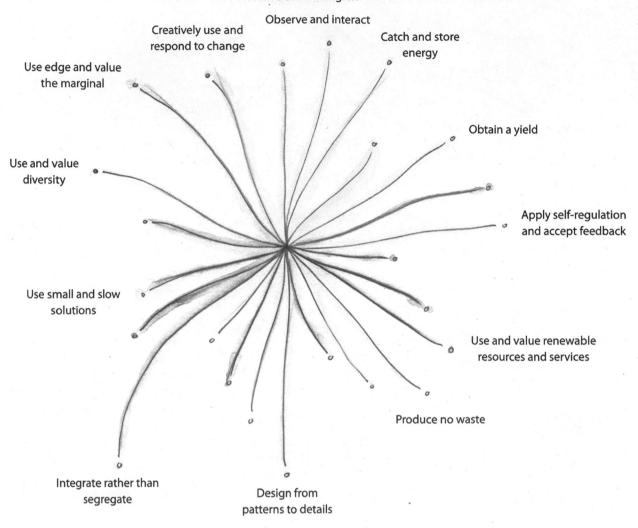

Observe and interact

Creatively use and
respond to change

Catch and store
energy

Use edge and value
the marginal

Obtain a yield

Use and value
diversity

Apply self-regulation
and accept feedback

Use small and slow
solutions

Use and value renewable
resources and services

Produce no waste

Integrate rather than
segregate

Design from
patterns to details

Farmers can improve production with design alongside the other managerial hats we wear. Design management (as a model for the design process) has helped us innovate because it emphasizes that design isn't stagnant, and so leaves us the creative head space to say, "this could be better." It reminds us to focus on universal factors like time/space/energy and consider how they balance out in our designs.

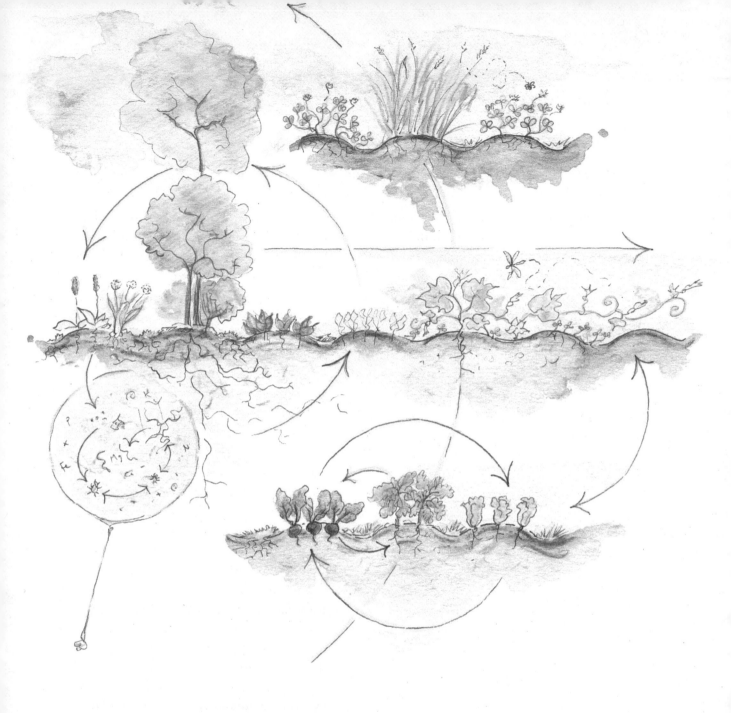

The Permabed System

The Permabed System

CONCEPT ORIGIN

The permabed system is the culmination of the previous sections in this book. It is an organized land patterning of permanent raised beds, based on whole farm mapping for ecosystem services and provides a framework to adopt and adapt agro-ecological production designs. It is designed to simplify diverse permaculture production for market gardeners. It is meant to serve you, your land, and your community, holistically.

"Where did the concept come from?" my dad asked.

I sipped my tea.

Original permabed concept designs

"Honestly, from being overwhelmed in the spring. From looking at large undifferentiated fields that had to be prepped, sowed and maintained with the diversity of species we currently grow and knowing there was something more."

I wanted a framework to build on, producing food profitably and ecologically, integrating the many innovations that are available to farmers, and make it easier to transition to a better form of market gardening.

Market gardening is like weaving a tapestry of crops across the land. Why should we have to reweave it anew every single year? A master musician can play a song many, many times in a year, yet the farm only has one chance each season to get it right. Maybe 40 chances in a lifetime! We need systems that allow us to build upon the previous year and design for continuity of farm ecosystem services through the seasons.

I wanted more intimate knowledge of the land, requiring a system that facilitated tracking all variables of the field: soil types, moisture regimes, weed pressure, previous season pathogen distribution, etc. I sought a framework for organizing site-appropriate

Patterned production above and below helping grow community.

perennial integration into an annual garden. I wanted principles to follow that could ensure movement toward agro-ecology. I grew intimate with the land — mapping, observing, sampling and recording.

I began to build up beds and stop breaking them down, to layer the soil more naturally and feed nutrients from the top down and bottom up simultaneously.

I decided to put sections of my land into permabeds and give them a specific place in

space. Next, I sought to develop a pattern that would allow me to confidently integrate perennials and annuals in a fluid system where neither would interfere with the other's efficient management and both would benefit.

I wanted to eventually achieve a woodland market garden, but one that was profitable and efficient for all scales of agriculture, where equipment could be used or it could be hand-managed, not exclusive to any one system, so the whole would be resilient for future farmers.

The permabeds became the organizing unit for a soil-building production. "Beds

are made, reformed, but never destroyed" became our motto. These beds could be linked up in an organized pattern across the land with groupings of beds (triads) regularly patterned as placeholders for the gradual introduction of perennials and annuals rotated in crop guilds between them. This means continued annual profit while building perennial investment. The permabed system would help us transition to profit resilience.

Permabeds come with a slew of benefits.

PERMABED BASICS

The permabed system can be adapted to many scales. The key principles are invariable whether you are operating on ¼ acre, 2 acres or 10+ acres. There are some techniques and tools that will change with operation scale, but the core is the same.

> *The middle ground is where we actually live. It is where we — all of us, in our different places and ways — make our homes.*
>
> — William Cronon

The permabed principles (page 183) exist for sustainable market gardening and movement toward profit resilience, where the farm is profitable because it actively supports ecosystem services which support sustained profit.

1. The permabed system uses garden environment mapping (GEM) to better understand the garden environment: soil, microclimate, ecology etc. It designates garden zones for improved annual production and best site selection for perennials. This also facilitates organizing crop rotations around better spring, summer, fall and near, middle, far spatial management.

2. It is fundamentally an organizational land patterning based on permanent raised beds that are reformed but never destroyed. These are patterned in triads, plots and blocks for management of diversified guild crop services and integration of crops and animals.

3. It is an ecosystem approach to market gardening that emphasizes relationships, cycles and services through investing in the soil, guild crop rotation and permission of succession.

> *The permabed system is a way of investing in the future by patterning agricultural diversity efficiently so it doesn't interfere with annual production while propagating agro-ecologies.*

> *We need an environmental ethic that will tell us as much about using nature as about not using it.*
>
> — William Cronon

Upslope

Ent.

Scutch grass creep

Dryest

Increasing stoniness

Wettest

Spring wet spot

Loam

Grey clay

Clay loam

Potential springs

Downslope

Permission of succession means we design our market gardens to allow natural succession toward an agro-ecosystem. Integrating perennials provides a natural ecological succession of microperennial ecosystems with an annual production. Reducing soil disturbance with permabeds means the soil ecosystem can evolve more naturally, building upon each season's growth.

What Are Permabeds?

Raised

The beds are raised above the normal grade of the land to create a productive growing space that is better drained, warms quickly and is less susceptible to compaction. This can be anywhere from 4 inches to over 12 inches.

It all starts with one bed.

Bed's have neighbors. Say "Hi!"

3 beds, the magic number. This is a triad!

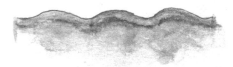

Triads can be linked together to form Permaplots (4 triads, one of which is for perennials).

Now we can see the foundation for a pattern. Beds & triads & plots repeat across the field.

ANA triads are three annual beds, and PERA triads are three perennial beds.

Do You Know?

Patterned Propagation

With patterned propagation we can bring about a transition to the desired agro-ecology bed by bed.

Plants from one plot can start a PERA triad somewhere else.

Plants may spread naturally, like raspberries or thyme, and be divided for replanting.

Plants are lifted and moved outwards

Plants are lifted and moved to new triads

Permabeds are are like weaving a tapestry, you need to start with the stitch and work out the pattern.

Triad: Three adjacent beds that are managed together and work toward symbiosis between current, prior and future crops. Using the Rule of Three we have made the triad the basic grouping of beds for garden patterning and management.

Do You Know?

Permabeds (permanent agro-ecological beds) may be less permanent than the name implies- since evolution, succession and change occur though soil organisms, perennial species and even the farm business production- yet this change is grounded in the beds' permanent placement (reformed and not destroyed) where abiotic and biotic systems can build upon previous seasons.

Permanence, a Place in Space

Permabeds are permanent: the form isn't destroyed nor the location lost by field-scale plowing, discing and cultivation. Each bed has a unique location.

**Beds are made and reformed, but never destroyed.
This permanence has many benefits.**

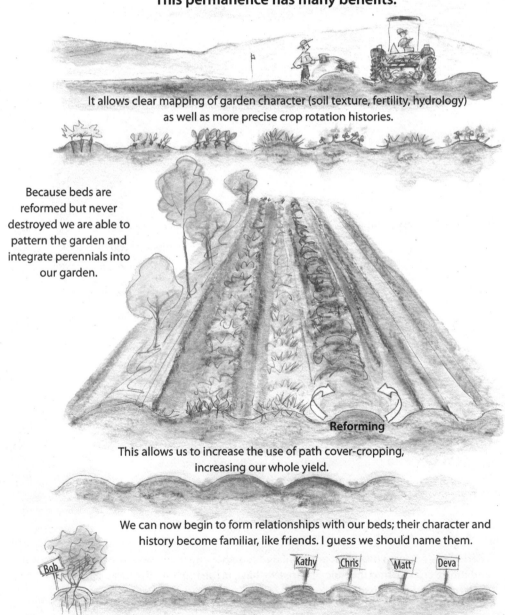

It allows clear mapping of garden character (soil texture, fertility, hydrology) as well as more precise crop rotation histories.

Because beds are reformed but never destroyed we are able to pattern the garden and integrate perennials into our garden.

Reforming

This allows us to increase the use of path cover-cropping, increasing our whole yield.

We can now begin to form relationships with our beds; their character and history become familiar, like friends. I guess we should name them.

Bob   Kathy   Chris   Matt   Deva

Worked Individually

Permanence prevents the mixing of disease and weeds through the entire garden, from field-scale cultivation.

Environmental Character Is Mapped

Permanence helps us better understand a bed's unique biotic and abiotic factors and records its crop history. We can place crops in beds based on their ecology, microclimate and seasonal production condition. This is mapped with garden environment mapping to create production zones.

Reformed not Destroyed

Beds are never destroyed, they are reformed by moving path material onto the bed top.

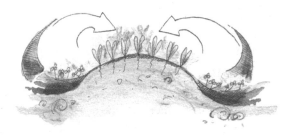

Uniform Architecture

The bed's width, length and shape are standardized. This allows us to work all beds with the same equipment scale, maintain crop production record consistency and facilitate crop rotation. We avoid field-scale equipment in conjunction with bed-scale equipment, which adds extra costs, in favor of a simple scale: bed-scale.

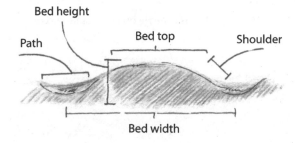

GREEN THUMB TECHNIQUE

Permabeds prevent bottlenecks in garden production because they accommodate exceptions in your crop rotation. Sometimes beds get too weedy, requiring seeding to cover crop, and new land must be found to grow the intended weed-sensitive crop. Permabeds can be chosen and put into service more easily than open, less-differentiated garden space.

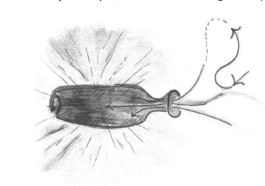

Principle — **Cover Crop Priority and In situ Composting**

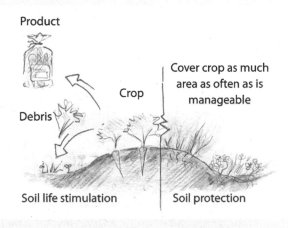

Permabed Architecture

Permabeds can be
made in many sizes

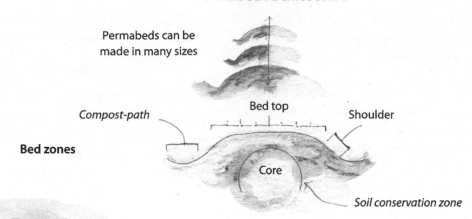

Compost-path   Bed top   Shoulder

Bed zones

Core

Soil conservation zone

Specific uniform architecture
Bed standardization entails a
chosen specific uniform
architecture to work with your
equipment, supplies and
methods.

Sandwich bedding will improve height

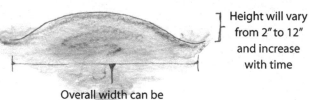

Height will vary
from 2" to 12"
and increase
with time

Overall width can be
set to match your
equipment

Alternate patterning

Alternate beds with
early and late maturing
crops is basic
permabed pattern

The weed management
zone = 1 bed top + 1
path

Offset mowers can
mow bed and one
path

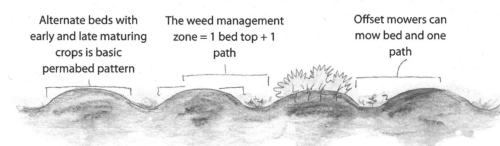

**Triad
organization**

Beds are organized into groups of three.
These triads form the next permabed pattern and the
basis of the guild crop production (see page 221)

All equipment is useable on every standard bed

**Remember: Sandwich bedding is the seasonal cycle of permabed formation — alternating cover
crops, crops and reforming. It can involve a bed-building period of self-sowing cover crops that are
routinely reformed.**

Permabed System
Dynamic & Efficient

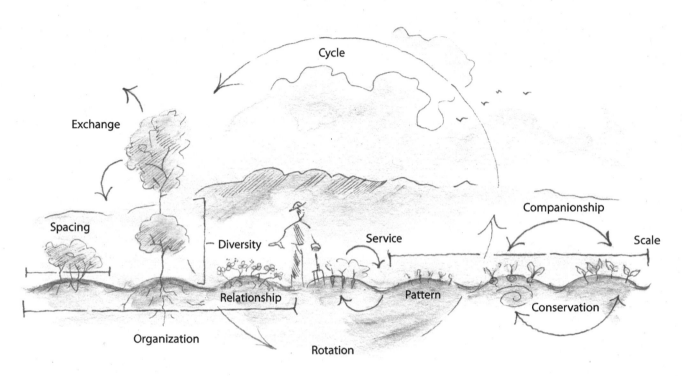

Efficient management and designed profit

Permabed Zones

1. Operation zone
* Over-bed space
* Bed top
* Shoulders

2. Accumulation zone
* Compost paths
* Cover cropped paths

3. Soil conservation core
* Protects soil life
* Isn't disturbed
* Allows soil life recolonization of operation and accumulation zone

Do You Know? Permabeds can be zoned like a biosphere, with an undisturbed core, a buffer and transition zone. These transition zones are key to success, for we cannot conserve the core without sustained profitability around it. Make the core the source of wealth and draw from the edges. Understand and value that the core *is the source of wealth* and avoid resource exploitation!

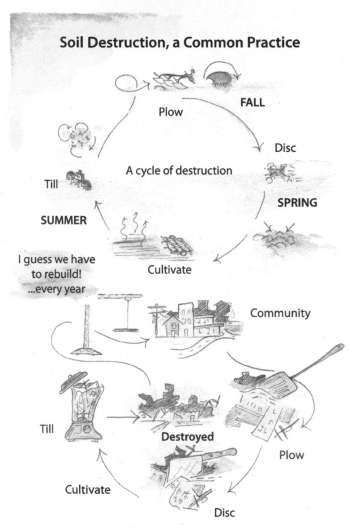

Soil Destruction, a Common Practice

FALL
Plow
Disc

A cycle of destruction

Till
SPRING

SUMMER

I guess we have to rebuild! ...every year

Cultivate

Community

Till
Destroyed
Plow

Cultivate
Disc

The common practice of aggressive tillage is akin to taking a city and flipping it (plowing), smashing it (discing), raking it out (cultivating) and blending it (tilling) and then asking the

* Soil organisms (people),
* Soil aggregates (buildings),
* Nutrient, water and air cycles (electrical, plumbing and road grids),

to function as usual for the production of our crops that absolutely depend on them!

Bed Reforming and Soil Life Conservation

No plowing

Beds are rebuilt

Soil life conservation zone

Soil life core

Soil life core

Permabed Zones are Like a Biosphere Reserve

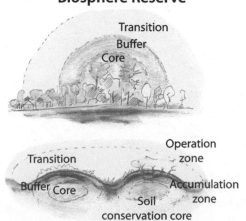

Transition
Buffer
Core

Transition

Buffer   Core

Operation zone

Accumulation zone

Soil conservation core

Benefits

1. Bed permanence allows fine-tuned management.
- Bed permanence allows you to fine-tune garden environment mapping, improve crop record keeping and focus on site-specific planting. This allows bed-by-bed land improvement, such as fertility or weed pressure amelioration, because the garden beds have this long-term presence.
- This is important for perennials placement and finding suitable garden conditions for annuals (like loose acidic soils for potatoes).

Permabed System Benefits

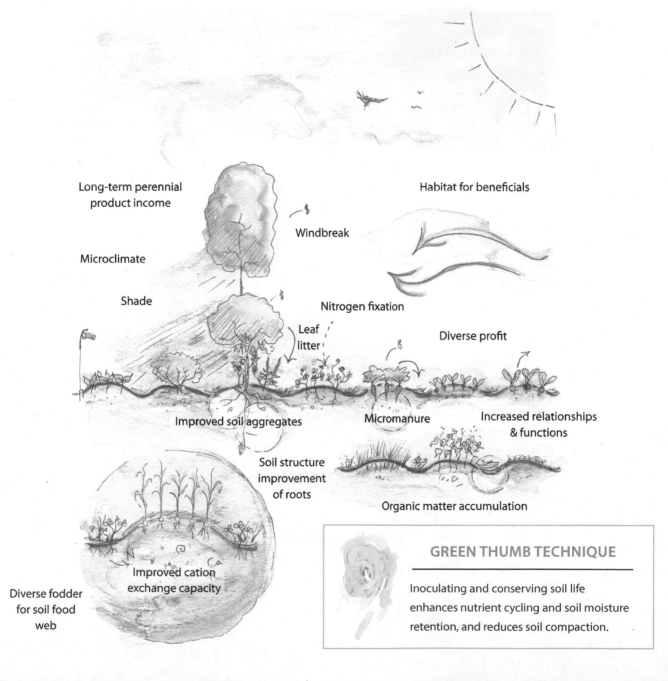

Long-term perennial product income

Habitat for beneficials

Windbreak

Microclimate

Shade

Nitrogen fixation

Leaf litter

Diverse profit

Improved soil aggregates

Micromanure

Increased relationships & functions

Soil structure improvement of roots

Organic matter accumulation

Improved cation exchange capacity

Diverse fodder for soil food web

GREEN THUMB TECHNIQUE

Inoculating and conserving soil life enhances nutrient cycling and soil moisture retention, and reduces soil compaction.

- Make full use of unique garden environments.

2. Bed standardization simplifies field operations.

- All equipment can fit all beds, instead of field- and bed-scale equipment, you have only bed scale equipment, less maintenance, storage and upfront cost.
- A single piece of bed-scaled equipment performs multiple functions. For instance, a flail mower can mow cover crops, be used to regenerate a raspberry bed or maintain an alley with one pass over uniform beds.

3. Healthy soil and soil life improves yields.

- Low tillage, bed reforming, and tarp culture replaces plow, disc and till operational cycles.
- This protects soil life in an untouched soil conservation zone.
- Permanent path ground cover builds soil integrity, meaning it improves aggregated structure, and enhances nutrient cycling and overall soil resilience.
- Soil conservation zone and path ground cover provide continuous habitat, so soil organisms can recolonize the bed's production zone following operational disturbances.

Permabeds Can Be Operated at Different Scales

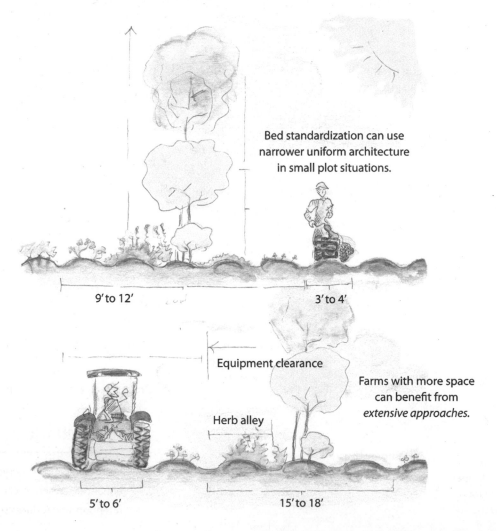

Bed standardization can use narrower uniform architecture in small plot situations.

9′ to 12′

3′ to 4′

Equipment clearance

Farms with more space can benefit from *extensive approaches.*

Herb alley

5′ to 6′

15′ to 18′

4. Early production

- Raised beds drain better, allowing earlier cropping.
- It is easy to see which beds are wet and dry and map this phenomena seasonally.
- Use any beds that are best for spring production because all the beds are already made.

5. Gradually benefit from perennials

- Perennial site selection and spacing is managed easily with garden patterning.
- PERA triads are placeholders for gradual diversification and layering.

Seasonal Benefits of Integrated Production

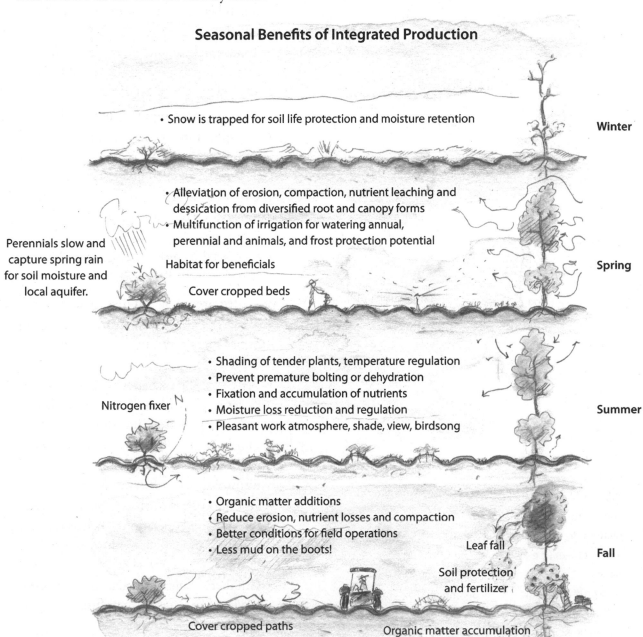

- Snow is trapped for soil life protection and moisture retention

Winter

- Alleviation of erosion, compaction, nutrient leaching and dessication from diversified root and canopy forms
- Multifunction of irrigation for watering annual, perennial and animals, and frost protection potential

Habitat for beneficials

Perennials slow and capture spring rain for soil moisture and local aquifer.

Cover cropped beds

Spring

- Shading of tender plants, temperature regulation
- Prevent premature bolting or dehydration
- Fixation and accumulation of nutrients
- Moisture loss reduction and regulation
- Pleasant work atmosphere, shade, view, birdsong

Nitrogen fixer

Summer

- Organic matter additions
- Reduce erosion, nutrient losses and compaction
- Better conditions for field operations
- Less mud on the boots!

Leaf fall

Soil protection and fertilizer

Fall

Cover cropped paths

Organic matter accumulation from leaf fall

6. **Extended flexibility of production**
*   Beds can be grouped and patterned in many ways.
*   Crop rotations can be easily adjusted if some beds need to be fallowed.
*   Any triad can be excluded from the rotation and put into a specialty purpose: say nursery or research production or trap crops.

Principle

Extended Flexibility

Design for unknown and future possibilities, for many different uses

Because permabeds have a fixed place in space, they can be managed in many ways without interfering with the production around them. If gardens are plowed on a field-scale each fall, then unique productions (like perennial beds) become an interference.

Diverse and unique productions are often a logistical nightmare. Permabeds provide a platform for umbrella management (see page 214), guild crop production and patterned diversity integration.

Fallow

Black tarp bed prep

In situ mulch production

Cover cropped access bed

Nursery

Green manure

Reforming

Many different perennial opportunities

7. **Maximizes whole garden yields through integrated production**
*   Operational excesses in the garden (irrigation overflow, manure splatter, debris residues) are better utilized with diversity. For instance, we alternate cover crops and crops in some garden plots and achieve much better productivity from both together.
*   Trees, chickens and cover crops integrated into the garden can improve soil integrity and enhance productivity.

8. **More garden inputs generated in situ**
*   Compost is made in field through crop-cover cropping (see page 219), compost paths and perennial pruning debris.
*   Other chosen enterprises (such as small livestock) would also generate fertility inputs in situ.
*   Coppiced hedges can produce trellis stakes in a chosen bed.

9. **Integrated pest and disease management**
*   Diseased debris is left at the same location of production through in situ composting (page 186).
*   Beds are not broken down each year through cycles of plow-disc-till, which allows field-scale equipment to drag disease around.
*   Disease and pests are reduced through bed reforming that buries disease without disturbing soil conservation core.
*   Pest life cycle disturbance is encouraged through field operations, such as late fall chiseling and fall tarp culture to break dormancy.
*   Guild crop rotation enhances companion services like deterring, confusing and trapping pests.

10. Increased access

- Standard beds allow tractors or carts to move freely through fields.
- Alternate maturity patterning (see page 185) produces a plethora of evenly spaced cover cropped beds to be designated as alleys when need arises.

11. Accrue ecosystem services and yields

- Improved soil fertility and cycling to offset fertilizer costs.
- Improved predator/prey dynamics for disease and pest control through beneficial insects.
- Improved garden conditions by crops for crops with guild crop rotation (see page 235)
- Gardens, when patterned for natural succession, receive the inherent benefits of long-term ecological complexity, intricate mycorrhizal networks, diverse nutrient cycling of soil organisms and micro-climate moderation of emerging treed agro-ecologies.

12. Organization improves management

- Permaplots provide organization framework.
- Three scale management: bed, triad, permaplot.
- Umbrella management of triads and eventual design of crop guilds.

13. Improved crop rotation

- Crop guilds are rotated instead of crops.
- Guild unit triads (see page 226) can be planted and seeded as needed.

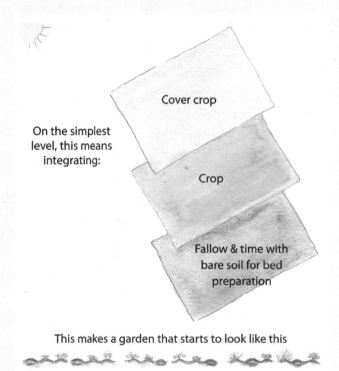

Principle

Integrate Productions
By integrating our current productions, we can achieve increased benefits between them.

On the simplest level, this means integrating:

Cover crop

Crop

Fallow & time with bare soil for bed preparation

This makes a garden that starts to look like this

- Ratios of crops within a guild unit triad can be organized for specific yearly demands.

14. Ease of crop experimentation/research

- Patterning for extended flexibility prioritizes research
- ANA or PERA triads can be allocated as research triads to trial new varieties or experiment with management techniques.

A triad of annuals is abbreviated ANA, and one of perennials is considered a PERA triad. These are patterned into a garden patterning as placeholders for that type of production at a ration of 3 ANA triads for every 1 PERA triad.

15. **Integration of productions results in more functional sharing of tools and supplies**
- Insect netting from early cucurbits can become bird-netting on dwarf cherry hedge.
- Irrigation for early mixed vegetables can also provide frost protection for tender fruit blossoms.

16. **Sharing of space and layering of functions**
- Better use of unoccupied space like open beds between squash that will spread or access space beside a row of fruit trees, both of which can be put into early maturing crops or cover crops to further benefit the whole system.
- Early asparagus beds can serve as apple tree alley come fall.
- Greens can be grown in between maturing zucchini.

17. **Easily adaptable and scalable for market gardens**
- Because the permabed system is based on uniform garden beds, common in market garden systems, it is easily adaptable.

Adherence to principles of production is paramount, to maximize ecosystem services!

- Management can be adapted to current equipment. The key is building a standard pattern for your scale, long-term production goals, and within environmental constraints and opportunity.

18. **Long-term profit potential and savings**
- Gradual integration of diversity opens new possibilities and benefits.
- There is diversification of potential long-term profit from new products, supply savings and efficiencies.
- Consider genetic resources like seeds, cuttings for grafting and regenerative suckers (like raspberries), or the savings of improved soil drought resilience.

19. **Enhances life style flexibility**
- Your profit centers can easily evolve to meet your changing life style.
- A growing family may mean you wish to move away from market sales and toward an on-farm you-pick forest market garden.
- The permabed system is flexible for production, so lends itself to changes in how the farmer wants to live and farm.

20. **The permabed system is increasingly beautiful and pleasant.**
- Shady rows here, song birds there, flower fragrances and tasty snacking berries

Permabed Principles

These are the permabed principles. They keep us focused on our goal of profit resilience in agro-ecological production. Benefits from following this system include:

- Crop-cover crops used for soil improvement, shade/windbreak and habitat.

- Nutrient cycling to crops by healthy soil food web.
- Use of overwinter trap crops (like kale)

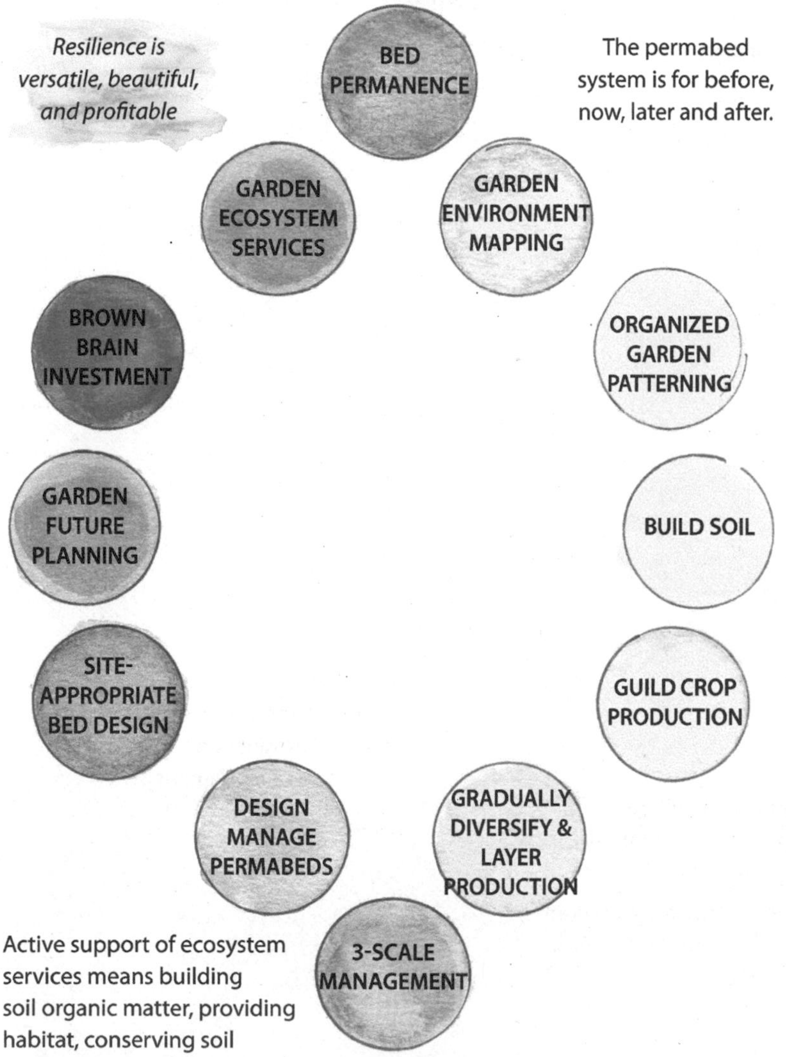

Resilience is versatile, beautiful, and profitable

BED PERMANENCE

GARDEN ECOSYSTEM SERVICES

GARDEN ENVIRONMENT MAPPING

The permabed system is for before, now, later and after.

BROWN BRAIN INVESTMENT

ORGANIZED GARDEN PATTERNING

GARDEN FUTURE PLANNING

BUILD SOIL

SITE-APPROPRIATE BED DESIGN

GUILD CROP PRODUCTION

DESIGN MANAGE PERMABEDS

GRADUALLY DIVERSIFY & LAYER PRODUCTION

3-SCALE MANAGEMENT

Active support of ecosystem services means building soil organic matter, providing habitat, conserving soil aggregation — it means designing systems that do this as part of the production routine.

Ecosystem Services

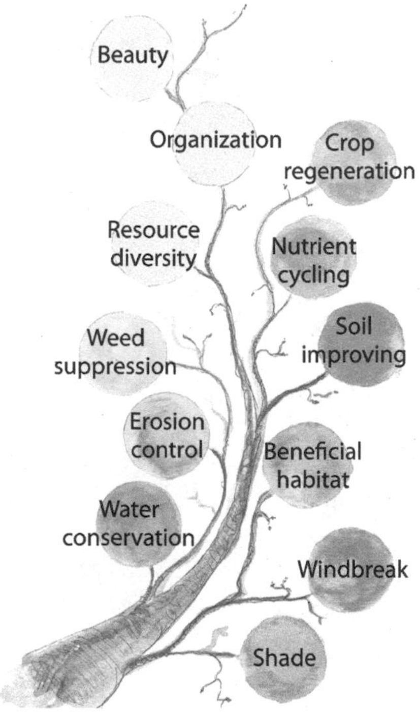

Beauty

Organization

Crop regeneration

Resource diversity

Nutrient cycling

Weed suppression

Soil improving

Erosion control

Beneficial habitat

Water conservation

Windbreak

Shade

We need to value these functioning systems and support them. We wouldn't expect our mechanic to repair the car for free. We wouldn't expect the stove to run without gas. We wouldn't assume to walk without feet.

The ecosystem services that are already essential to farming can provide us with so much more — once we give them space to function and support them through *best management practices*.

Bed Permanence

Permabeds have a fixed place in space, calling card of their own, allowing us to know their environment and crop history better. They are reformed on an annual or biannual basis. Beds are never destroyed only reformed. Their path material (often cover cropped) is lifted onto the bed top as compost. This process allows the common practice of plowing in disease and debris without destruction of soil and loss of organized space.

Garden Environment Mapping

Garden environment mapping is the description and tallying of the garden's environmental character (resources, climates, flows, etc.).

- *Bed soil profile:* structure, fertility, organic matter, microclimates
- Soil food web (soil life)
- Water movement, drainage, holding capacity
- Bed site, slope, stoniness and bedrock depth
- Aspect, microclimate
- Garden zone designation
- Site-appropriate planting (see page 197)
- Seasonal observation
- Mapping using garden map templates

DESIGN TIP

Your farm is within a regional climate and your garden situated in a microclimate. Map climate for species that require this range of conditions.

Bed Permanence

Diversity requires a patternable design for management

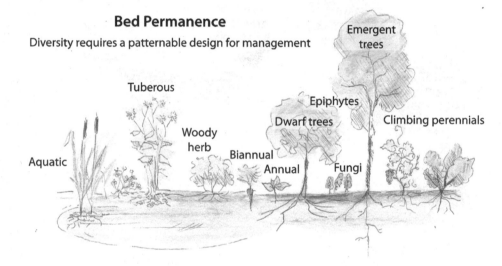

Tuberous

Emergent trees

Epiphytes

Dwarf trees

Climbing perennials

Woody herb

Biannual

Aquatic

Annual

Fungi

Create patterns for integrating layers of productions to allow best management

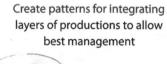

Use of permabeds provides uniform definable space for patterning of annuals, perrenials and integration of livestock.

Permabeds also allow us to easily integrate layers overtime without interfering with current production.

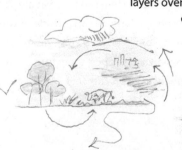

Organized Garden Patterning

This is the repetition of permanent beds across the land and their organization into triads and permaplots (organizational units) for the purposeful patterning of integrated production.

Examples of organized garden patterning include:

- Permaplot patterning: the alternating of perennials and annuals across the permaplot in a set ratio
- Alternate maturity patterning (AMP): the alternating of crops by days-to-maturity (DTM) within triads
- Permaplot enterprise patterning: the alternating of enterprise focus between permaplots (vegetables, orchard, livestock)

Principle

Organized garden patterning is the alternating of diversity based on species, maturity and intended production goals. It includes the grouping of beds into triads (group of three beds) and designating them for annuals (ANA) or perennial (PERA) production and patterning them in a 3:1 triad ratio across the garden. PERA triads are placeholders for perennials until they are planted as such. Because permabeds hold a distinct place in the field and the garden is managed with bed-scale equipment, ANA triad operations can occur around PERA triad production, and the permanent nature of perennials need not interfere with annual crop producitivty. Both productions benefit from the other, receiving microclimate cooling, organic mulches and improved habitats for soil organisms and beneficial insects.

Garden Patterning

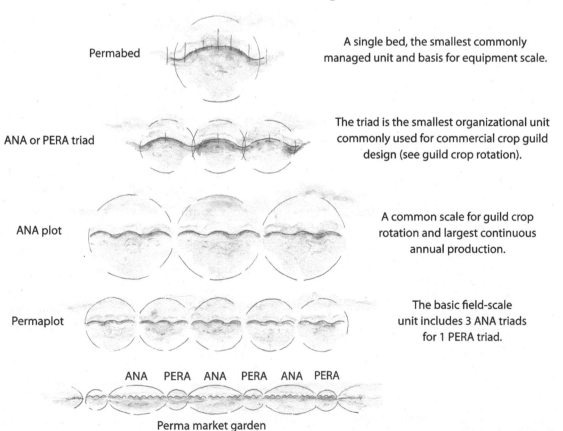

Permabed — A single bed, the smallest commonly managed unit and basis for equipment scale.

ANA or PERA triad — The triad is the smallest organizational unit commonly used for commercial crop guild design (see guild crop rotation).

ANA plot — A common scale for guild crop rotation and largest continuous annual production.

Permaplot — The basic field-scale unit includes 3 ANA triads for 1 PERA triad.

ANA   PERA   ANA   PERA   ANA   PERA

Perma market garden

Bed reforming

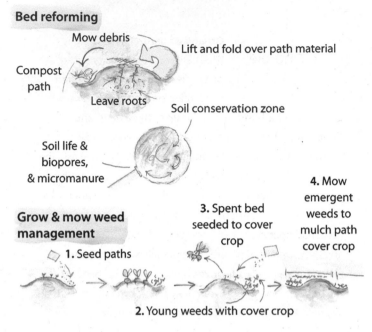

Mow debris

Lift and fold over path material

Compost path

Leave roots

Soil conservation zone

Soil life & biopores, & micromanure

Grow & mow weed management

1. Seed paths

2. Young weeds with cover crop

3. Spent bed seeded to cover crop

4. Mow emergent weeds to mulch path cover crop

Compost Paths for Under 1-acre Operations

Nutrient-rich debris

Cover crop fixing nitrogen

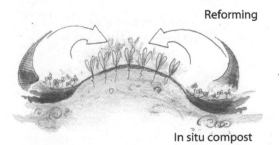

Reforming

In situ compost

Build Soil

(We will elaborate on these in more detail)
- Bed reforming
- Cover crop priority (relay, path, intercrop), most of the time
- Crop-cover crops,
- Soil conservation core
- In situ compost
- Patterning to minimize bare soil
- Mow and grow weed management
- Analyze the full soil profile (1 m deep) for best management of soil resources
- S is for SOIL: Work the soil strategically: seldom, softly, slowly, shallowly — and stratified

In Situ Composting

In situ composting can be *in-bed* and with *compost paths*. These techniques build compost right where it is needed with the disease and pest vectors buried in the bed. Compare this to removing your debris to compost piles where (usually poorly maintained) the disease festers and its movement is less well understood and controlled.

In situ compost examples:

- Leaving crop debris in situ and reforming the bed over top.
- Cover cropping into the debris of a spent bed.
- Pulling debris and leaving it in the path if a clean bed top is needed for new crop.

Do You Know?

Permabed records can include disease, pest and weed history, fertility, carbon, inoculants and amendments.

- Allocate beds for compost production. This can include hugelkultur beds, where pruned branches can be laid in furrows and recovered, cover-cropped as a long-term carbon enrichment and biological resorts.

Cover Crop Priority

- We prioritize minimal gaps of time when the soil is bare. Use of intercropping and relay cropping helps us to do so. It is paramount that we stop a system of bare soil farming, a till and kill system for removal of weeds, preparation of seedbeds and avoidance of disease.
- Beds are mostly in either crop or cover crop.
- Sheet mulching to fry weeds is preferred over prolonged cultivation.
- Paths have a permanent cover crop to compete with weeds in this edge environment (white clover, annual rye, oats, fescue).
- Beds are sown to cover crops as soon as crops are removed or undersown where suitable.
- Sow your summer beds into cover crop immediately.
- Sow your fall beds into cover crop for spring crops.
- Sow your late fall crops into cover crop in spring.
- Manage paths for maximum cover crop opportunities.
- Try to maximize your cover crops benefits by allowing time for them to develop.

Principle   **Cover Crop Priority**

Relay cover crops
(see glossary)

Path cover crops

Nitrogen-fixation,
root penetration, organic matter
accumulation

- Always have cover of some sort.
- Crop → green manure → manure → cover crop → compost → crop
- Aim for 50% of land in cover crop at any one time.

Example: We can differentiate where across a field the pH increases or decreases and plan beds with lower pH for potato rotation.

We can find beds of best drainage and include them in rotation for early crops that need dry ground in April.

A spent bed is one where a crop has been grown and harvested yet there may still be services, such as mowing the debris over broadcasted cover crop seed as a mulch, or allowing it to overmature as a habitat hotel for beneficial organisms to be nurtured.

Guild Crop Production

(See section on guild crop production)

- Focuses on crop services to greater garden ecosystem, organized crop companionship, and guild crop rotation

  Makes use of:
- Ecological mimicry in guild design
- Critical crop matrix analysis: a comparative chart that analyzes crops for their companionability
- Crop guild assembler: a design template

Gradually Diversify and Layer Production

Production occurs on many vertical layers in a forest. We should mimic this strategy in the market garden to maximize photosynthetic yield. Garden patterning helps us allocate space for trees, shrubs, vines, etc. into the garden.

We can further diversify by cover cropping paths, establishing pollinator hedges and designing treed farm lanes, as well as integrating chickens, bees or other enterprises, according to your guild enterprise production.

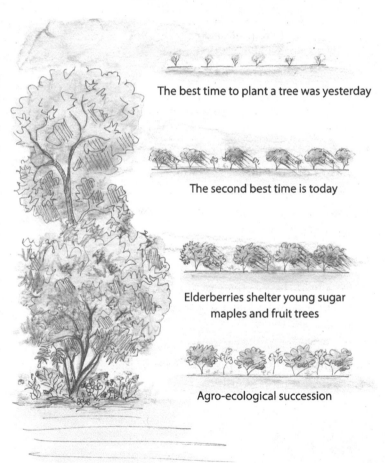

The best time to plant a tree was yesterday

The second best time is today

Elderberries shelter young sugar maples and fruit trees

Agro-ecological succession

Do You Know?

There are many benefits to integrating production. These include:

- Minimizing large tracts of bare soil vulnerable to erosion, compaction and nutrient loss.
- Enhanced habitat and inoculation sources for more diverse and healthy populations of soil organisms.
- Maximizing potential companionship between crops, cover crops and perennials, such as pest deterrents or shading/cooling of summer greens.

Elderberry syrup, a treat for any market

Laneway production makes management a routine

Integrate through organized garden patterning

3-Scale Garden Management

We manage on three scales:

- Permaplot (9 beds) used for guild crop rotation and permaplot enterprise patterning
- Triad (3 beds of annuals or perennials): used for building crop guilds, and organizing field operations like tarp culture and row cover

- Bed scale (1 bed and paths) is the only scale we purchase equipment for.
- Bed scale is further managed on three levels:
  - Rows: weeding, drip, seeding
  - Shoulder: hoeing, hilling
  - Path: cover crop, composting, reforming

3-Scale Management

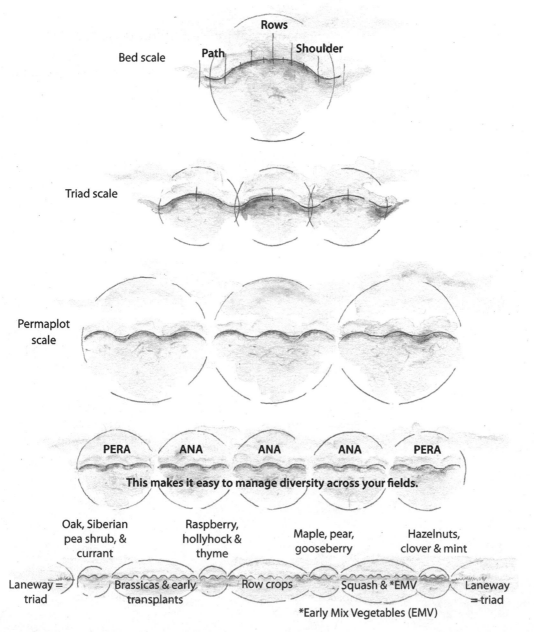

Bed scale — Rows, Path, Shoulder

Triad scale

Permaplot scale

PERA   ANA   ANA   ANA   PERA

This makes it easy to manage diversity across your fields.

Oak, Siberian pea shrub, & currant

Raspberry, hollyhock & thyme

Maple, pear, gooseberry

Hazelnuts, clover & mint

Laneway = triad   Brassicas & early transplants   Row crops   Squash & *EMV   Laneway = triad

*Early Mix Vegetables (EMV)

GREEN THUMB TECHNIQUE

Mulch the ends of PERA triads. This is an edge that should be left unmanaged. Mulch them with hay mulch, cover crop them or cover them with a weed barrier and use them as a space to set supply crates.

Design Management for Permabeds

Permabeds must be design managed to best meet your site and situation, and your goals now and into the future. They should follow the principles for design management, achieving better space/time/energy productivity.

Principle

Extended Flexibility

Bed permanence and a well-organized garden patterning permit flexibility for diverse garden uses now and into the future. It is possible to put any bed, triad or plot into other productions or other purposes. Including:

- Specialty crop beds
- Research triads
- Small livestock rotations
- Pig aerator
- Cover-cropped access alleys
- Crop-specific edge beds
- Hoop house triads

False sowing is the practice of preparing a fine seed bed for germinating weed seeds for shallow cultivation prior to seeding crop in the now clean bed.

Design Management for Better Production Cycles

Design management in a permabed system adopts and adapts production cycles that can build soil and improve ecoystem services. Permabeds have given us the liberty to experiment with different options and observe.

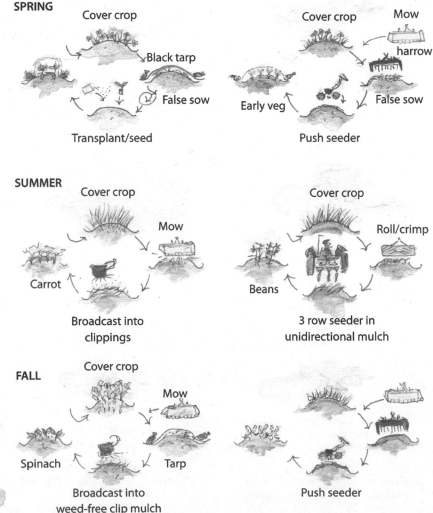

Site-appropriate Design

Beds should be designed and organized for the specific site. Decisions such as bed architecture, contouring, layout and patterning are infuenced by:

- climate, topography and property size
- drainage, moisture regime, solar aspect.

The goal is the beds work for your land and how you want to produce on it.

Garden Future Planning

- Gardens are designed for their utility into the future.
- They should be beautiful and useful.
- The garden pattern should be versatile for unintended agricultural practices.
- On larger acreage design should allow for mechanized harvest/regeneration potential so the production isn't limited to small tools.
- Wider lane possibility is another way to ensure space is versatile for future needs.

Principle

Consider current and future harvest flexibility

There are many ways to harvest crops.

Make sure your designs are flexible for harvest, this makes or breaks efficiency. Consider the options and harvest requirements of both ANA and PERA crops.

Many perennial shrubs can be machine harvested, and it is important that we leave room/ design for this option. I strongly believe any operation should grow machine-harvestable perennials. Perennial production on farms should be designed for flexibility so future decision-makers can adapt the plantings, increasing or decreasing scale and management style.

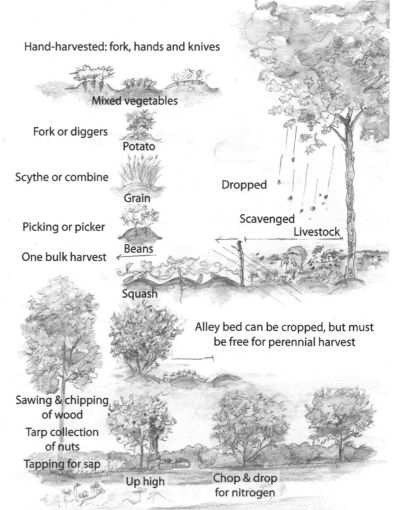

Hand-harvested: fork, hands and knives

Mixed vegetables

Fork or diggers

Potato

Scythe or combine

Grain

Picking or picker

Beans

One bulk harvest

Squash

Dropped

Scavenged

Livestock

Alley bed can be cropped, but must be free for perennial harvest

Sawing & chipping of wood

Tarp collection of nuts

Tapping for sap

Up high

Chop & drop for nitrogen

Future farmers, may want to manage perennial production differently, and design should allow them to scale up or down easily, mechanizing or not.

Brown Brain Investment

(See Holistic Planning section)

- Investing in ecosystem services
- Investing in alternative approaches
- Investing in soil annually!

S Is for SOIL

Seldom work the soil and do so in advance, using methods that are less aggressive

We prepare spring beds the summer prior to production using soil improving green manures followed by bed reforming and a winter cover crop.

Basic Permabed Cycle

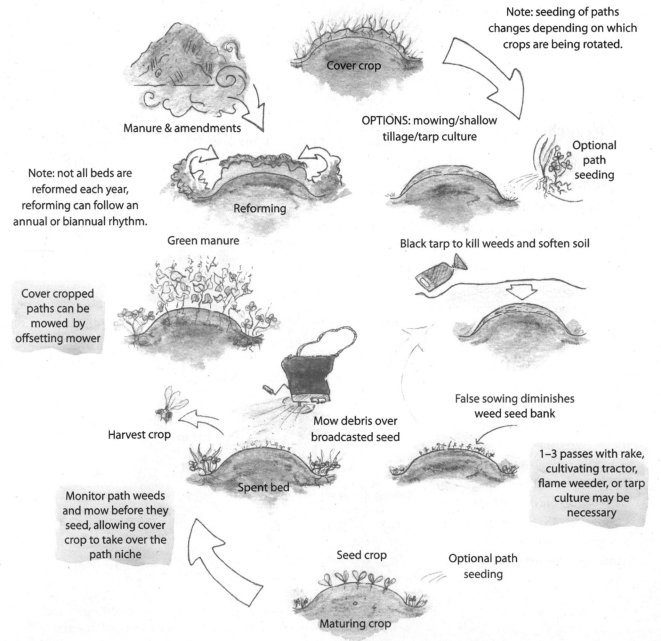

Manure & amendments

Cover crop

Note: seeding of paths changes depending on which crops are being rotated.

Note: not all beds are reformed each year, reforming can follow an annual or biannual rhythm.

Reforming

OPTIONS: mowing/shallow tillage/tarp culture

Optional path seeding

Green manure

Black tarp to kill weeds and soften soil

Cover cropped paths can be mowed by offsetting mower

Mow debris over broadcasted seed

Harvest crop

False sowing diminishes weed seed bank

Monitor path weeds and mow before they seed, allowing cover crop to take over the path niche

Spent bed

1–3 passes with rake, cultivating tractor, flame weeder, or tarp culture may be necessary

Seed crop

Optional path seeding

Maturing crop

GREEN THUMB TECHNIQUE

Mow broccoli tops, leave roots in place, and cover crop on reformed beds.

- Soft and slow methods of improving soil for crops are best.
- Cover crops like oilseed radish improve structure by creating biopores (opening in soil left from decayed roots and tunnel organisms).
- Black sheet mulching for killing weeds instead of quick plow and discing.
- Shallow tillage is more than enough to produce good soil/seed contact for germinating fine seeded crops.
- Set the depth gauge on bed preparation equipment to prevent working deeper than 2" to 6" (deeper if transplanting).
- Stratified beds use equipment that doesn't invert the soil profile. Soil is developed from the top down. We need to mimic this phenomena to build the best quality soils.
- Use of a permanent bed system that emphasizes bed reforming, cover cropping, sheet mulching and shallow tillage ensures the profile isn't destroyed but enhanced.

Topics for Consideration
Efficient Garden Management
Many multidisciplinary management concepts should be remember and applied in a permabed system to maximize efficiency.
- Rule of three
- Uniformity of beds
- Standardization of practices
- Umbrella management
- Alternate maturity patterning

The Golden Triad
It occurred to me when designing the permabed system that the number three is the first to have complex interactions. So why go any further? Simple is best. Use the number three to organize everything. Gardens are organized into triads both horizontally and vertically.

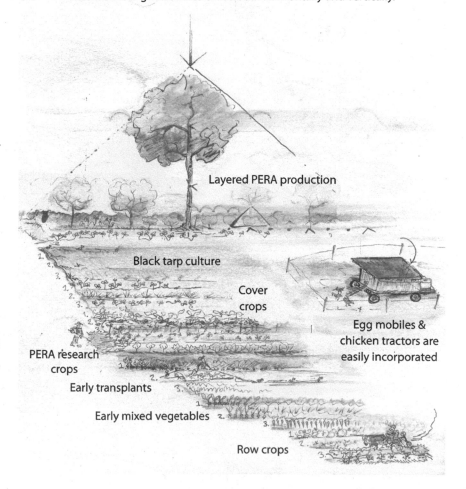

Layered PERA production

Black tarp culture

Cover crops

Egg mobiles & chicken tractors are easily incorporated

PERA research crops

Early transplants

Early mixed vegetables

Row crops

- Gaining perspective
- Value space/time/energy
- Kit organization
- Color-coded field to-do stakes

Scalable Perspective

Perspective at different scales can help with various solutions. Take a chartered plane over your farm, climb a hill, lie on your belly amidst your crops and use a microscope to examine soil life. Visit other markets as a customer, visit other farms as an observer, and spend time in other professions. Solutions come when we look from different heights, angles and layers.

Build Local Knowledge and Find Global Solutions

Specific knowledge of your local area, whether neighborly wisdom or specific ecological phenomena can enhance your farm management. Draw on the experiences available globally. Many similar circumstances have been encountered, and there is a wealth of tools.

Economy of Motion

How many motions does it take to complete a specific task? How many stages to complete the operation? Eliminate extra work! Count your actions, whether transplanting broccoli or bunching carrots (Pull carrot, put elastic around once, twice, break top off halfway, lay in row). Sometimes you'll find unnecessary motions, like switching the bunch unnecessarily from one hand to another. "Economy of motion" — thanks Eliot Coleman!

Umbrella Management

We group crops according to similarity of management and use efficient mechanized solutions for some cash crops and intensive hands-on solutions for others. Make note of similarities between crops: small seeded vs. large seeded, many-rowed crops vs. easy hilling of single-row crops, succession often vs. a few successions.

Understand and Value Crop Life Cycles and Services

Crops have their own life cycle, which reveal many hidden yields and services. For example, crop-cover crops, when crops are allowed to bolt, producing more organic matter, insect habitat and nectar sources. These crop-cover crops can be undersown to rye or clover and cover cropped after.

→ Manage for crop life cycle services to maximize their benefits. For instance, the thick-rooted crops such as broccoli and kale can improve the soil. We never want to pull another broccoli stalk again and designed our rotation to benefit from biopores created during Brassica root decomposition.

GREEN THUMB TECHNIQUE

Introduce, Grow and Conserve Beneficial Organisms, Profit from Their Valuable Interactions

There are many beneficial organisms like nitrogen-fixing bacterial, mycorrhizal fungi and parasitoid wasps.

Four-Season Cover

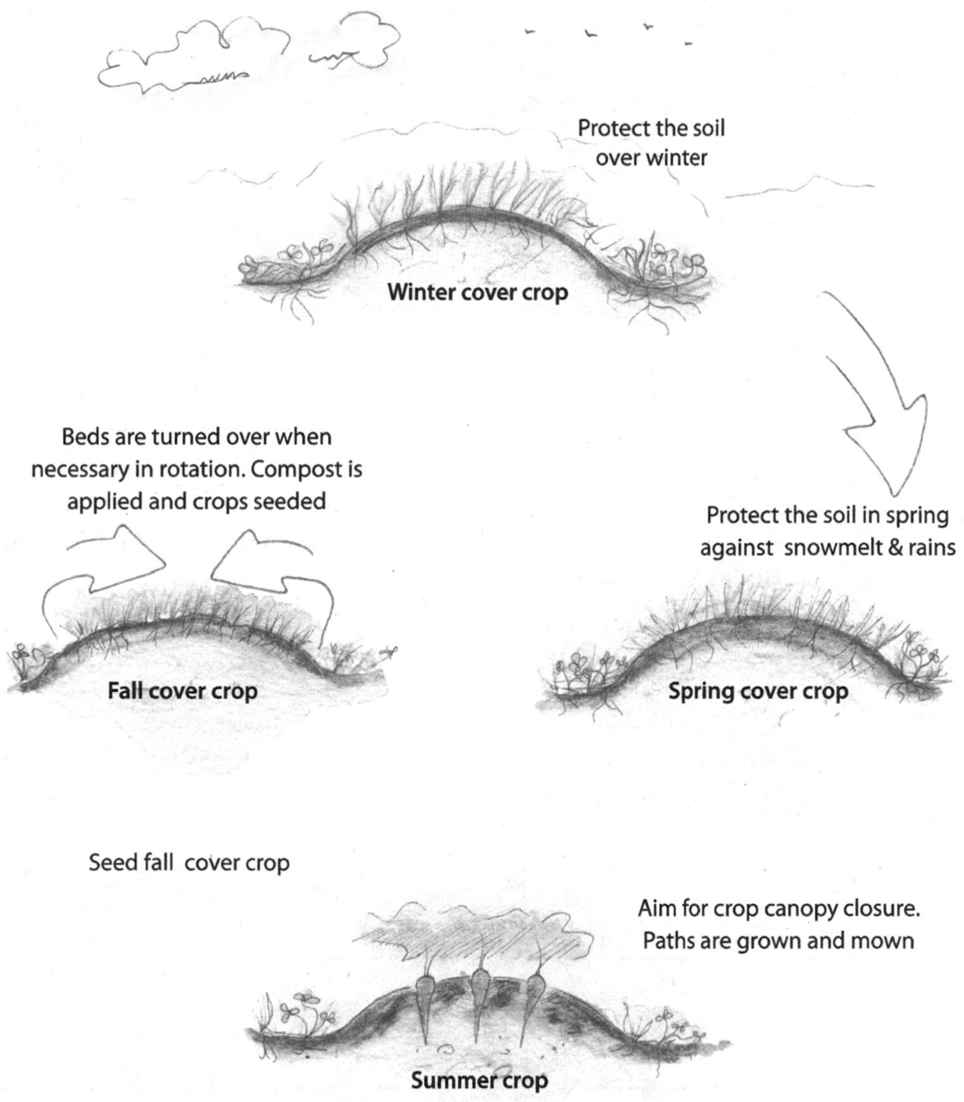

Protect the soil
over winter

Winter cover crop

Beds are turned over when
necessary in rotation. Compost is
applied and crops seeded

Fall cover crop

Protect the soil in spring
against snowmelt & rains

Spring cover crop

Seed fall cover crop

Aim for crop canopy closure.
Paths are grown and mown

Summer crop

There should always be something growing in your beds.
You can double or triple crop your garden or grow more cover crop.

*Understanding natural rhythms
will simply help manage the
complexity.*

PLANNED LAND PREPARATION

This is important because its focus is meeting the needs of production (weeding, fertilizing, seed bed formation) in advance, before you are growing. This can be 3 years in advance, 3 months or 3 weeks, its about being prepared. Working ahead uses less resources, time and energy

Six Planning Stages for Land Preparation

1. Garden area assessment and mapping

The whole farm mapping process may reveal potential garden areas. These areas should mapped in more detail for their ecological character, environment and crop history.

This includes mapping soil, water, vegetation, frost pockets, etc.

Garden environment maps allow gardeners to observe, note and map differing garden conditions. First map the whole garden as areas of wetness, stoniness, etc., then refine environment records on a plot, triad and bed scale. Map on graph paper, in Excel or using templates.

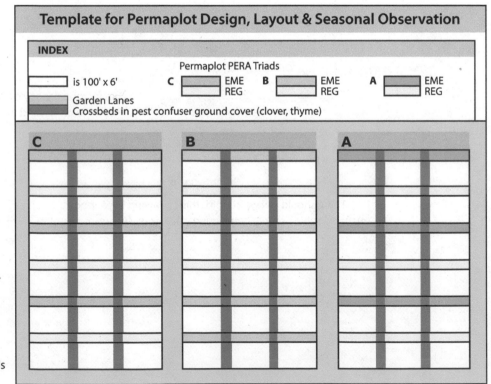

NOTE: I make these much larger than displayed so data can be input into cells. Usually one cell is 25' x 6'.

*EME is Emergent PERA triads
*REG is Regenerative PERA triads (see page 208) for definitions

Do You Know?

Ecosystem Approach

An ecosystem approach to market gardening requires more detailed understanding of the garden area. This helps the placement of perennial crops in the correct conditions needed for multiyear health and productivity.

Perennials have an important role to play in ecosystem services, and so their proper placement is a paramount reason for garden environment mapping.

Also, this mapping reveals opportunities for annual production by where land is more suitable for a gin crop. For instance potatoes could be rotated the soil is sandier and/or lower pH, as is their want.

Note: I use the terms garden character and garden environment interchangeably. I like the word *character* as it denotes familiarity and personality to the garden. This is, after all, what we want to do — see the garden environment, specifically each bed, as having a character of its own — and respect it as a friend and business partner.

Garden Environmental Character Permanence

When mapping, it is helpful to know which features can be amended and those which are best left and used for their best potential, page 93 talks more about this.

Unchangeable character
- Elevation/aspect
- Parent material
- Rockiness

Last-minute action rarely provides long-term satisfaction!

- Stoniness: quantity and size of free stones
- The presence of rock outcrops and/or depth to bedrock

Changeable character
- Wetness (can be difficult in some cases)
- Top soil
- Vegetation, grasses, shrubs,
- Fertility, pH
- Organic matter content
- Soil life quantity and quality
- Shading

Garden Frequency Zonation

We organize crops into garden frequency zones based on the frequency of their harvest and the intensity of care. We call these near, middle, and far frequency zones.

- Near: Kitchen garden, demonstration gardens, experimental gardens, index guilds, hugelkultur compost areas
- Near: Nursery, greenhouse, hoop houses, small plot trials
- Middle: Picking crops: peas, beans and zucchini
- Middle: Main season crops: lettuce, beets and early potatoes
- Far: Single-harvest storage crops: squash, pumpkins and garlic
- Far: Hay fields

Frequency Zonation

Near   Nursery, hoop houses

Middle   Main crops

Far   Storage crops

Garden Seasonal Production Zonation

- Some zones of a garden are determined to be best for spring, summer or fall productions and scheduled for this.
- For instance areas that are usually drier are better for spring production.
- Yet, areas that are often wet in spring may be workable in summer and have better summer moisture.

Upslope

Scutch grass creep

Dryest

Increasing stoniness

Wettest

Acidic

Spring wet spot

Alkaline

Loam

Grey clay

Clay loam

Potential springs

Downslope

GREEN THUMB TECHNIQUE

Passive Garden Improvement Techniques include strategies to enrich soil, eliminate weeds and improve land preparation without as much labor, equipment and energy. As we know space/time/energy is a balance and often these techniques require more space and/or time but save a lot of energy.

- Tarp culture: using large black tarps to fry weeds
- Self-sowing cover crops are good in beds you want to enrich for a few years
- Sandwich Bedding (see page 202)

It is always easiest to apply a good kick-start of nutrients, get rid of thoroughly of noxious weeds and design a quality layout from the get-go.

2. Field preparation (3 years before you start cropping the land)

This includes:

- Overall layout of fields for garden, define edges, lanes, plots
- Tree, shrub removal and/or selection for use in garden intercropping
- A great time to get large animals on the land
  - Manure
  - Get goats eating the shrubs
- Find, set contours and decide on water-flow management
- If possible, open more land than you need and cover seed green manure crops, like rye and then red clover, to improve soil until land is needed.
  - It is always better to open land and just let it improve slowly; when you need it, it is ready.
  - Consider opening 3 times the land you currently need (rule of three).
- If limited for land, put extra care into weed and fertility management so you can triple crop!
- This is the time for large-scale fertility amendment and cover cropping.
- If you are working full-time to budget for farming, now is a good time to really focus your energy on passive garden improvement techniques.

3. Initial bed preparation (season before production)

- Layout of permaplots
- Seed garden lanes into hardy fescues and clover
- Build permabeds and lay out crossbeds (seed into perennial ground covers)
- Continue to build bed soil structure with sandwich beds
- Introduce local micro-biology with local woodland soils, compost teas, manures, etc.
- General micronutrient amendments as per needs
- Inoculate cover crops for improved nitrogen fixation
- Can begin planting PERA triads

4. Preproduction bed reforming

(we do this every year or two, where it fits into rotation)

- Reforming
- Preproduction false sowing (use tarp culture, basket weeding etc.)
- Refined fertilizing includes amendments specific to your crop rotation.
- Winter cover crop: prewinter stabilizing and bed protection

5. In-season bed preparation

- Cover crop killing/incorporation
- In-season bed preparation (power harrow, tiller, roller/crimper, etc.)
- Review "Destroy Your Soil 101" (page 57 to 59) to improve tillage management for in-season bed preparation.
- In-season false sowing: use row cover to stimulate weed germination to speed up preproduction weeding.

6. Crop rotation preparation

This is focused on how we manage beds in relation to the next crop to be grown.

- Relay cover cropping
- Crop-cover cropping
- Crop rotation managed for preceding crop
- Dealing with debris, roots, mulch
- Guild crop production

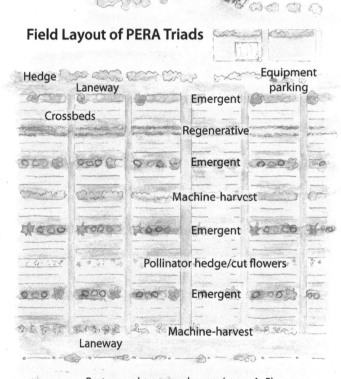

Field Layout of PERA Triads

Hedge
Laneway
Crossbeds
Emergent
Regenerative
Emergent
Machine-harvest
Emergent
Pollinator hedge/cut flowers
Emergent
Machine-harvest
Laneway
Equipment parking

Pasture or hay or cash crop (zone 4–5)
Another form of Alternate Patterning is the switch between types of perennials that occupy PERA Triads.

Being hasty with aggressive weeds will haunt you like the Hydra did Odysseus. Thistles and quack grass will come back vigorously from the tilled pieces that have been dragged throughout your garden. This is very bad in a perennial/annual hybrid system!

Do It Right from the Start

Weed management: Remove the current vegetation, especially perennial grasses and other invasive weeds. You will rebuild biodiversity once you have selected the species that work best for your land.

Existing trees: You can build your permabeds around any wild species that fall into your patterning. But I would advice against leaving species: too much unorganized wild areas within a garden design in favor of an organized garden patterning of alternating PERA and ANA triads.

Fertility: Apply copious fertility in June and seed buckwheat cover crop, red clover and rye. Better to stimulate your soil life and build organic matter through nutrients taken up by cover crops.

Layout the laneways around the garden and build all your raised beds, and design for crossbeds in longer gardens.

Contour: Put your beds roughly on contour; for steeper areas, contour is much more important. Full contour will reduce efficiency of some mechanized operation. Research more on planning contours, keyline and other work being done in this area of land management.

A decision delayed is a decision better made: Rushing into any project without the proper design, planning and research can result in costly mistakes. Examples are placement of a building in a poor location, a bed width that doesn't make sense for the future scale of your operation, a sprinkler system that is too powerful and causes soil compaction, etc.

GREEN THUMB TECHNIQUE

Working and Farming

It is easy to arrange for manure spreading and cover cropping while maintaining a different job elsewhere. Allow your soil to work for you while you make money and plans for your permaculture market garden. Opening land, adding manure and seeding a good cover crop and leaving it for a year or three is a very productive move toward starting your farm.

CREATING PERMABED LANDSCAPES

Permabeds from Scratch

Old fields have many perennials. To prepare a market garden, we must remove these competitive species.

Measure the area to plow. Plowing turns the sod under, burying the leaves so they can't photosynthesize and lifting the roots so they dry out.

Discing will break up the furrows and allow you to cultivate.

Another option is covering the area with poly tarps. You can buy "hay tarp" at farm supply stores. This method is great as it minimizes tractor passes. The annuals and perennials fry. Leave the tarp on the ground for one full year. Next summer · a quick discing and cultivation is needed before cover cropping.

Cultivating twice a week in May and June will eliminate the perennials by dragging them to the surface, knocking any soil off the roots and exposing them to the sun. Broadcast some oats and peas in August before the rain to cover crop for winter. If you choose to use poly tarps, then just leave it on.

Seed the field in buckwheat by midsummer. You can broadcast the seed before a rainy week or seed it with tractor. Apply manure before cover cropping. Weed seeds in manure are smothered by buckwheat.

Flail mow or disc buckwheat when it flowers. Broadcast winter rye with red clover in late August. This cover crop protects soil through winter. The following summer, make sure to mow

the rye during the *milk seed* stage this is when the seed is soft and gooey.

Once the rye is dead, the clover will come up through it, fixing nitrogen and provide an excellent cover over the winter.

Now it is time to use some bed-forming equipment. This can be done with a tractor and disc bedder or with a walk behind tractor with rotary plow.

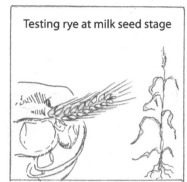

Testing rye at milk seed stage

Permabeds from Scratch: The Basics

After site assessment and selection, it is time for land preparation.

A good time to begin rotating crops, small livestock and intensifing cover crops.

1. Old field
This is an ideal time for large animal in situ fertilizing

Specific amendments

Permabeds

Plowing is best for larger plots

Black tarp best for small plot

Disc
2. Field preparation

Cultivate

4. Preproduction bed preparation

Compost

Bed forming

Manure

Buckwheat

3. Initial bed preparation

Rye

Clover

Lay of the Beds on the Land

Lay out your beds in relation to slope. I use rough contours and end up with beds of even widths. This is essential for management on our scale.

Sandwich Beds

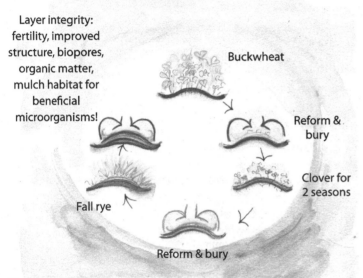

Layer integrity: fertility, improved structure, biopores, organic matter, mulch habitat for beneficial microorganisms!

Buckwheat

Reform & bury

Clover for 2 seasons

Reform & bury

Fall rye

Benefits of Sandwich Beds
• Cost-effective way of building healthy permabeds
• Can be done in future fields as field preparation
• Can be managed while working another job! Let nature work for you while you business-plan and save money.

Results:
• Accumulation of decomposing debris layers stabalize bed shape and integrity
• Building soil structure, biopores, organic matter and fertility
• Weed seed bank reduced, outcompeted by cover crops
• Selective seed bank can be established, flush them out when needed.
• Steady-state regeneration bed is necessary for crops

GREEN THUMB TECHNIQUE

Build 3, 6 or 12 permabeds now! Seed them in a mix of cover crops and let them self-sow untill the time is ripe to work them into you rotation. You'll thank yourself.

If you are 100% on contour, you will never get even beds. That is fine for production of grains with perennials. But it is not good for market gardens. Market gardens are one of the most diverse and intensive agricultural productions and require standard bed widths for efficient management. Gradual curving of the beds is fine, but bed width must be similar.

My favorite thing to see is a bed in perpetual cover crop waiting for me to plant it. If you think you can't "afford space" for this, think about the costs of poor bed quality. In our case, areas of heavy clay, where harvesting root vegetables is backbreaking, are being prioritized for sandwich beds.

To Make Permabeds with Tractors

Start at one end of your field and make a run following your markers, dropping your disc bedder at the same start and end points. Turn around at the end and place your left wheel (if working right to left down a field) directly into the left wheel tread from the previous bed. I prefer wider paths for cover cropping and place my wheel halfway onto the fresh-cut edge my disc bedder left. Chiseling or sub-soiling beforehand will loosen subsoil and break up any plow pan (compaction layers).

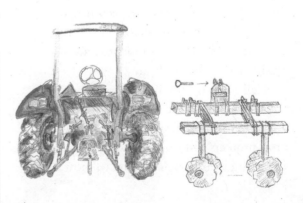

For walk-behind tractors, I suggest much the same process but using more passes with a rotary plow or swivel rotary plow. Consider the work done by Jean Martin Fortier and other operations working with solely walk-behind tractors. A taut string line, two stakes and can of spray paint can help get your garden plot off to a straight run.

Now it is a good time to apply compost to the bed tops with a spreader or by hand. You

Currently we only use our walk-behind in greenhouses

Rotary plow helps make raised beds

can also drive up the bed with a tractor with the loader bucket full and shovel it out. Seed your beds down into cover crops and use tarp culture when you are ready for production of crops to kill and decompose them.

Beds and Land Contours
There are three ways to build permabeds in relation to land contours. Each has benefits and drawbacks.

On Contour
• Beds follow the contour lines of the land.

Everybody in the countryside has field-scale equipment: don't buy it! Rent it or, even better, pay your neighbor to do primary tillage (plowing, discing, etc.)

• All points on a bed are of equal elevation, as such beds are not of equal width and they may not be of desired length.
• Very important on steep slopes.
• On sloping land this will have maximum nutrient- and water-holding capacity and reduction of runoff/erosion.
• Could hold too much water in a very wet area or undesirable where early production is required.
• Will prevent erosion with cover cropping, and bare soil will suffer less localized erosion in heavy rains.

Roughly on Contour
• Beds are made perpendicular to the slope of the land.
• All points on the bed may be roughly of equal elevation; beds prioritized to be of equal width rather than equal elevation.
• This is a good choice in most gently sloping land for efficient garden operations, especially for production over 2 acres.
• On sloping land, this will have medium nutrient- and water-holding capacity and higher risk of runoff/erosion.
• May facilitate drainage in spring.

Remember when planning and purchasing that investments must make you more money than they cost. Pass decisions through the testing questions to make sure they are a good fit in your holistic plan.

- Could prevent erosion with use of cover crops, bare soil may suffer erosion. Mapping of vulnerable areas to maximize permanent cover in these areas.
- Designing water catchment and infiltration systems are useful
- Reduce overland flow into garden through integration of PERA buffer triads and upland earthworks to capture run-off.

Beds are More in Parallel with Slope

- All points of the bed are not of equal elevation; beds are of equal width.
- On sloping land, this will have the least nutrient- and water-holding capacity and increased risk of runoff/erosion.
- Could result in very quick drainage on sloping land.
- On sloping land, cover crops could be used to prevent erosion, but any bare soil would suffer severe erosion in heavy rains and snow melt.

Seasonal Observation and field demarcation of water movement can lead to other design possibilities.

Garden Environment Mapping Influences Layout and Water Management

PERA buffers reduce run-off into gardens

Reduce seepage from hillside

Water meadow laneway

Evaluate spring-fed pond

Do You Know? Keyline design can help field-scale agriculture conserve water and build soil. There are good articles available online.

Ways of dealing with seasonal water flow when not fully on contour

Water meadow laneways are useful for areas where the garden is only experiencing heavy seasonal flows. In this model, water flows its course along cover-cropped lanes that filter nutrients and slow water with soaker speed bumps and channel water toward a pond or drainage ditch. In summertime these slightly sloped laneways are available for traffic.

Hedgerows drainage ditches consist of a triad where the central bed is dug out and turned into a seasonal waterway. The material is applied to the outer beds which are then planted into water-tolerant perennials.

Capture upland water into reservoirs before it reaches lower fields, design pond, storage tanks, perennial berms, flood irrigation systems and other water-saving opportunities.

Please continue to research local systems for how to manage water. It is a fundamental element that is both useful and destructive.

Springs and Seasonal Water Tables

These should be turned into ponds, reservoirs and streams.

The worst erosion will be experienced on steep land with complex sloping. Precision contours and smaller-scale production is best!

Organized Garden Patterning

Before we broaden our discussion with guild crop rotation, lets review and expand upon its foundation: organized garden patterning

We pattern our land, gardens and beds to ensure there is an overarching harmony to our production, building seasonal and agro-ecological rhythms into our production. With patterns we can integrate trees into the garden, add animals and trial for research. Patterns give us the flexibility to efficiently manage diversity for profit resilience.

Organizational Land Patterning Review

Land patterning can take many forms. For example, we can organize fields into pasture and hay or cropland into divisible fields of rotated crops, using the divisions as windbreak, edible fencings and long-term profit opportunities. It can also include the layout of roads, lanes and paths and the inclusion of ponds, streams and reservoirs. It is how we fence, trellis and where we build sheds, barns and houses. Land patterning is the pattern we leave across our land. It is also a point of entry for making a pattern that is more versatile, multifunctional and productive for farms. We can choose to pattern our land on the contours of our hills, on the flows of water, wind and light and the depths of our soils.

I like to learn about things again and again and again, first for to see it, then to know it and lastly to understand it

Integrating Small Animals

A mobile hen hut can easily be rotated into a well-organized garden environment by fencing off a triad or an entire plot with portable electric fencing.

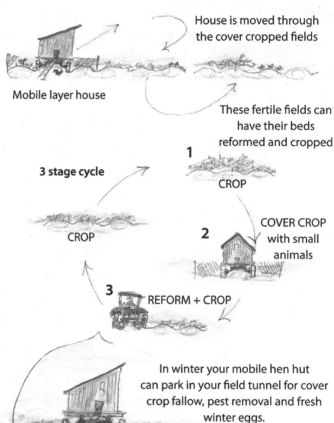

Mobile layer house

House is moved through the cover cropped fields

These fertile fields can have their beds reformed and cropped

3 stage cycle

1 CROP

2 COVER CROP with small animals

3 REFORM + CROP

CROP

In winter your mobile hen hut can park in your field tunnel for cover crop fallow, pest removal and fresh winter eggs.

Note: you must balance your need for new ground with your available covered cropped ground. Consider also moving them around perimeter lanes of permaplots, and/or have designated grazing grounds (perhaps land unsuitable for gardens).

Easy Organized Patterns for You

Here are a few patterns that are easy to integrate into the farm and should be considered essential in a well-managed farm.

- Plant your property lines in trees, evergreens to the north and west, with fruits and nuts to the south and east.
- Plant your driveways with sugar maples and nut trees, and your laneways with fruit trees.
- Plant along your water ways and ponds with fast growing species like willow, elderberry and birch.
- Then organize the rest of your property into a consistent pattern, roughly on contour, and integrate productions.
- Make it accessible, multifunctional and versatile for future production.

See things in other ways, hear them as other sounds and say them with other words

Hedges & living fences

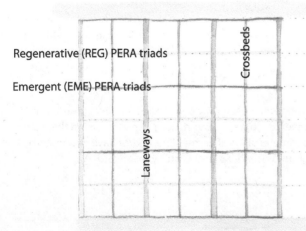

Regenerative (REG) PERA triads

Emergent (EME) PERA triads

Crossbeds

Laneways

*We will discuss the major groups of perennial triads on page 208

Repetition of an Organized Unit, Patterning the Organized Space and Rotation within the Framework

In order to pattern you must have:

- Repetition of organizational units: such as permabeds and triads.
- Then you can pattern by creating an ANA/PERA ratio, planting PERA triads and placing unique triads, such as hedgrerow drainage ditches, buffer beds beside laneways, etc.
- Followed by rhythm, the rotations and evolution of ecology over time and space.
- Rotate annuals, animals, etc., between permaplots and allow PERA triads to mature

Patterning and Layout

This is the division of the garden into three primary organizational units and three primary flow ways.

Organizational units:

- The bed is 6 feet wide × 300 feet long (or your standard).
- The triad is 3 beds managed together. There can be both ANA triads for annual production or PERA triads for perennial production.

Principle

Permabed Patterning for Integration

The permabed system is for the management of agro-ecological production.

- Integration of common productions (cover crop, crop and fallow)
- Integration of perennial and annual production
- Integration of animals and other productions

- The permaplot is 3 triads of rotated production and 1 triad of perennials. These triads could be in annuals, become a primary perennial production or be used for animals or other production. For instance, a meadow for honey.

Flow ways:

- Garden lanes occupy 1 triad of space that is given over as a roadway around or between plots.

- Alleys are cover cropped beds running parallel with the rest of the garden. They are part of the 12 beds in a permaplot, but they are dedicated for needed seasonal access, often part of a PERA triad.

- Crossbeds are covered-cropped beds running perpendicular as pollinator strips, access ways and space dividers.

Triad Terminology, Syntax and Patterns

When designing for ANA triad guilds it is important to clarify terminology.

- The crops in the middle bed are called center crops and the outer beds are outer crops.

   Describe a triad like this: outer/center/outer as lettuce/squash/lettuce.

   Describe a more complex triad like this: outer/center/outer as lettuce/squash/corn/beans/lettuce, where the center beds have three crops.

- When discussing a crop guild, the jargon consists of a *key crop*, which serves as a starting point for finding good *neighbors*.

- The key crop is the focus of production and compatible neighbors are found because they can be umbrella managed or are beneficial companions.

- The difference being: outer beds could have the key crop or the inner bed could be the key crop.

- This difference of key crop placement is part of variations of triad organization, called guild unit triads (GUT).

- Often, the key crop has the longest DTM (days to maturity), since it requires the longest management consideration. In a squash plot the squash are the key crop. In a plot of early mixed vegetables (salad/carrots/radish), the carrots are the key crop.

Select Your Guild Unit Triad

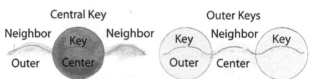

The key crop is the crop that is most important for management consideration.

Triad Organization

PERA Triads Bed Possibilities

Regenerative or machine harvest triads

Good neighbor

Alley crops

Main crop

Emergent & hand-harvest triads

This PERA triad has outer beds in companion crops to help mitigate pests

This PERA triad has a compliment of perennial species in the center bed and the outer beds are used as alley crops to access the key crop for harvest.

ANA beds are rotated between PERA triads

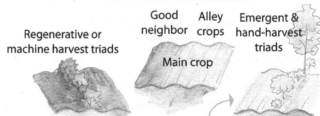

Elderberry

REG triad: Currant/thyme, elderberry, clover/chives

Depending on scale you can have 3 similar crop triads or 3 different

EME triad: Oak, maple & pear clover/oak, maple, pear/asparagus

Triad is organized around umbrella management

Triad is organized around companionship/symbiotic benefits

Triad is organized around alternate maturity patterning

Triads should strive for organization.

The acronyms help us conceptualize patterning and save space when using design templates (see page 188)

- ANA: Annual triad
- PERA: Perennial triad
- Permaplot: PERA/ANA/ANA/ANA

You can even name your plots after their PERA triad (red oak, apple, hazelnut) to create themed permaplots. With time you can evolve this pattern plot by plot and choose to have entire plots in perennials, pasture or other production.

- PERA/ANA/ANA/ANA/PERA

or

- OAK/SQUASH/SQUASH/SQUASH
- HAZEL/EMV/EMV/EMV/MAPLE

The first plot is called red oak permaplot and the second is hazelnut permaplot, which would be followed by sugar maple.

Types of Triads

- ANA triads: dedicated to ANnuAl production of crops as guilds or ANimAls
- PERA triads for PERenniAls, either
  - Regenerative triads (REG) for species that can be harvested or regenerated by machine, or
  - Emergent triads (EME) for species that can form an overstory canopy.

Principle

Management through Alternate Bed Patterning

Garden patterning that maintains that alternate beds are more alike than adjacent beds.

Basic pattern

Alternating FAST maturing and MODERATE maturing annuals

Space/time resource partitioning

Early crops harvested and late crops spread

Grow & mow weed management

With early crops harvested, it is easy to mow with equipment; you can offset the mower to clear the path too!

Relay cover cropping

Cover cropping spent beds

Make use of life cycle services

Early crops left to flower or for seed, later crops are harvested and cover cropped.

GREEN THUMB TECHNIQUE

Alternate Patterning

Alternate patterning is a unifying theme across permaplots. Alternation is such a fundamental principle of nature, like the peaks and valleys of sound waves that we use as a guiding unifying patterning within the garden.

Types of Patterning

- **Permaplot patterning:** ANA and PERA triads alternate across a field in a ratio of 3 to 1

- **Perennial field patterning:** PERA Emergent and PERA regenerative triads are alternated in PERA placeholders (designated PERA triads).

- **Alternative maturity patterning:** Beds within triads alternate by crop maturity (very important)
- **ANA guild unit patterning:** includes arrangements of triads into specific annual assemblies used within your rotations such as carrot/salad/beet. These have an objective of improved management and inter-crop services.
- **PERA guild unit patterning:**
  NOTE: PERA patterning refers to the specific grouping of species in a single PERA triad (raspberry/mixed fruits/asparagus). The objective is improved management and mutual crop services. This creates crop guilds and their various guild unit triads.

- **Bed/path patterning:** Alternating of bed crop / path / bed crop / path across the whole field. The point is that our paths can grow useful crops and provide fundamental services and should be seen as their own pattern.
- **Guild bed patterning:** *Guild bed patterning* refers to how you plant crops within a single bed. For ANA beds this includes planting more than one species in rows for a full length of the bed or planting block areas of 25 or 50 bed feet of one crop with the same bed feet of another crop. Remember additive planting (page 220), adding beneficial plants in transplant trays so they get planted out in a desired ratio.

Pattern the Land

Adopt, adapt and evolve as is needed

Mixing annuals with perennial islands (PERA Triads)

Permabeds are an intensive land organization.

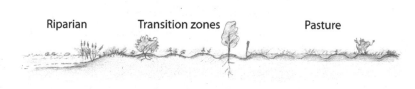

Annuals between establishing perennials

Riparian   Transition zones   Pasture

Even where I haven't made them, I think with their spacing

Patterned garden

Mixed agro-woodland

DESIGN TIP

Permaplot enterprise patterning is the alternating of enterprise focus between permaplots (vegetables, orchard, livestock). Entire permaplots can be assigned the role of an orchard or pastureland according to your guild enterprise production.

Choose your own ANA/PERA ratio if need be. Remember to adopt, adapt and evolve organized garden patterning for your goals, enterpises and environment.

PERA Triads

PERA triads are dedicated for perennial production, yet may include annuals or cover crops, either until perennials are planted or with the perennials until they mature. There are many possibilities within the emergent and regenerative categories.

PERA placeholder: the space in the permaplot meant for a PERA triad, which may be in annuals until the suitable species is researched, propagated and planted.

PERA Triad

There are many possible arrangements for PERA triads. All should consider harvest, access, and the habitat of species.

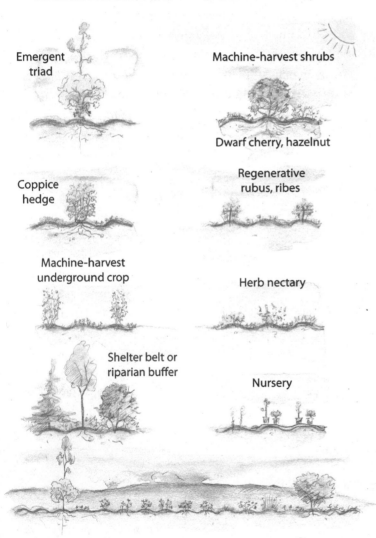

Emergent triad

Machine-harvest shrubs

Dwarf cherry, hazelnut

Coppice hedge

Regenerative rubus, ribes

Machine-harvest underground crop

Herb nectary

Shelter belt or riparian buffer

Nursery

PERA Species Selection and Placement

- Your decision for species selection should include your holistic goals and long-range guild enterprise production goals. Don't plant species you don't have market research for, or species that won't work with any long-range plans.

  Overall, PERA triad design emphasizes grouping species into guilds for inter-guild services and efficient umbrella management. Especially important are: harvest, weeding, pests and fertility management.

- Your perennial crops should be hardy enough for your climate and trialed in small amounts in property zones 1 or 2. This way you know they are hardy for field planting, you'll have propagation material and an index guild. The index guild (zone 1–2) gives a glimpse into the field production (zone 3–4) and is crucial for efficient perennial management while you also manage myriad annuals that simultaneously occupy your attention.

- Species selection should be based on site suitability and garden layout. Consider garden environment mapping and scrutinize the PERA placeholder triads for suitability of desired perennials. Also consider other perennial species that may match PERA placeholder conditions: moisture regime, soil type and microclimate.

- Lastly, maintain perennial field patterning, which is helpful in achieving a woodland agro-ecology. This provides a balance of woody perennials cores, herb edges and annual meadows. It will allow emergent PERA ecologies with nuts, pears, apples, plums, herbs and regenerative triads with elderberries, hazelnuts, raspberries and currants. This ecological mimicry provides long-term perennial production with open spaces managed for favourite annuals).

- A big concern is whether a tractor should be able to drive over the key bed or beside the key bed. The key bed is the middle production bed or both outer beds. This applies to all scales of production from 13hp walk-behind tractors to 4-wheel tractors. There are three access considerations.

GREEN THUMB TECHNIQUE

Here are some surprising harvests:

- Baby cabbages sprouted after main cabbage harvest
- Fennel fronds sprouting after main harvest
- Kale hearts, the inner most tender leaves of kale cut out just before the freeze-up

Garden Environment and Crop History Mapping

The garden environment is neither uniform nor stagnant. It is important to map and understand for PERA triad placement. Eventually ANA and PERA permabeds will have a crop history that can be recorded for improved rotation and evolution of intercropping designs.

Weed seed bank, perennial weed presence, localized weed preferences

Hydrology: drainage, wet spots, water flows, soil water-holding capacity

Crop history: what, where, when

Soil health: structure, aggregates, organic matter, soil organisms

Warmth, aspect, more solar gain

Fertility: what, when, how much

Select Your GUT

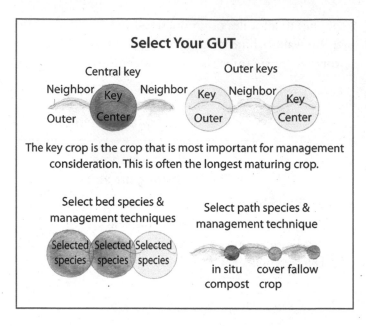

Central key

Neighbor ... Neighbor ... Neighbor

Key Center | Outer | Key | Outer | Neighbor | Key Center

Outer keys

The key crop is the crop that is most important for management consideration. This is often the longest maturing crop.

Select bed species & management techniques

Selected species | Selected species | Selected species

Select path species & management technique

in situ compost   cover fallow crop

1. Access for machine harvest. Crops like hazelnut, dwarf cherries and Saskatoon berries can be machine harvested.
2. Access for easy picking. Crops like apples and pears will benefit from your being able to drive adjacent and harvest off the back of a truck or have it nearby for heavy loads.
3. Ability to lift the bed for suckers or plants for nursery sales or propagation into other plots.

Patterning helps integrate agro-ecological services such as pest confusers/deterrents, cover-cropped alleys and beneficial insectaries, into our gardens in a flexible way.

Patterned Propagation Is Good Math

It includes *addition* of perennial crops, *division* of the expanding plants and their *multiplication* throughout the permaplots. The *SUM*: your agro-ecology

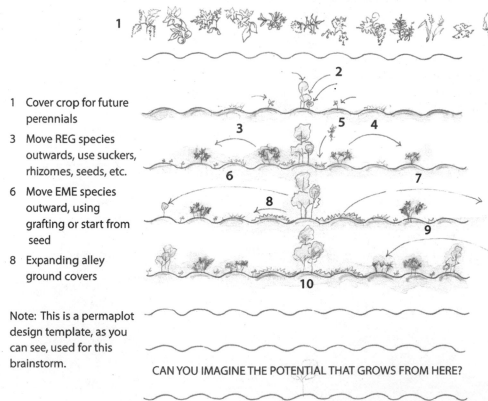

1 Cover crop for future perennials

3 Move REG species outwards, use suckers, rhizomes, seeds, etc.

6 Move EME species outward, using grafting or start from seed

8 Expanding alley ground covers

Note: This is a permaplot design template, as you can see, used for this brainstorm.

2 Initial trials of different species to fill different niches

4 Pioneer species

5 Add new species from nursery to fill changing niche, for instance shade tolerant ground cover

7 Use triad as a nursery for favorite species to lift and plant elsewhere or for sale.

9 Import an exciting new variety from research nursery

10 Pattern propagation has resulted in an affordable establishment of a full PERA permaplot

CAN YOU IMAGINE THE POTENTIAL THAT GROWS FROM HERE?

Primary PERA Triads

1. Emergent (EME), hand-harvest triad

- Characterized by pioneer trees like birch with fruit trees. These are sheltering species like oak, maple or chestnut, which will become the dominant over-story.
- Bed patterning can include berry bushes, herbs, and flowers in a garden's early succession stages, evolving toward more mature ecosystem production.
- Dominant species are hand-harvested: apples, chestnuts, lumber, fungi, grapes.
- REMEMBER: that all PERA triads have an evolving relationship with neighbor beds in order toserve as access (cover cropped alleys).
- These access beds can grow crops like thyme, mint or asparagus that won't interfere with driving over the bed with truck or tractor.

2. Regenerative (REG) triads

- Characterized by species that can be regenerated by mowing, reforming and heavy compost application.
- Species like raspberries, elderberries and currants can all be mowed to the ground, composted thickly and allowed to regenerate.
- Also, these species could be lifted for transplant sale or to be add to another similar bed elsewhere.
- Easily used for patterned propagation.

PERA Triads

Emergent triad
Mixed fruit, nut and berries are planted as EME triads as they mature, to hold the largest perennial species

Vines
Vines should be planted as an EME triad as they can easily climb up future intercropped trees.

Hedgerows
Pre-existing hedgerows can be incorporated into the permabed plan. Their edges can be modified with black tarps and planted to cover crops or shade-tolerant ground cover crops in order to prevent the spread of weeds and grasses.

GREEN THUMB TECHNIQUE

Hugelkultur compost beds

Once a PERA triad is maturing you may wish to prune. Create a specialty triad on the north side of this PERA guild. Furrow the beds with a potato furrower, fill them with the branches and cover them with your hillers.

This buried woody matter can help build bed integrity over time, similar to sandwich bedding, but with a woody core. Cover crop with red clover and oilseed radish for 3 years and then use tarp culture to prepare for planting.

This is a great slow release for anticipated expansion of you PERA triad through patterned propagation or a research triad.

REMEMBER: woody debris traps available nutrients for a period of time requiring added nitrogen to make in situ compost.

Requiring Machine Access

Access over or immediately adjacent to many PERA triads dictates their architecture.

Center access

Regenerative Triads

Crops like currants and raspberries can benefit from mowing, compost additions and even prescribed burning. Planted without intercropped trees allows mechanization of these processes: flail mow, flame weed, manure spread and let regenerate. In addition, canes could be lifted by a tractor for propagation.

Side access

Machine-harvest Shrubs

Crops like hazelnut and dwarf cherries have the potential to be machine harvested and should be given space to do so. Because perennials are long-term, it is better to pattern for more access (place in central or key bed) and make use of the adjacent beds with annuals or cover crops until they are needed for access.

Seasonal access

Protection & Nursery Combo

Low-growing, sun-loving crops can serve as a nice ground cover in other PERA triads. But the serious strawberry grower will want the air flow, sunlight and easy access for mulching, picking and bird netting. It is hard to lay netting around trees in a bed where strawberries are also a ground cover. Also, both strawberries and asparagus could be lifted and divided for propagation into new PERA triads when they don't share a bed with woody crops.

I don't recommend slash-and-burn (building stable soil carbon is much more important), but sometimes fire can be good to eliminate a build-up of disease.

3. Machine-harvested (MAC) triad

This is a subgroup of REG triads where regular machine harvest is specified in the design.

- Characterized by species that can be machine-harvested.
- This can be broken down into crops that require the tractor to drive over the key bed in the triad (sunchoke root harvest) or to drive adjacent to the key bed (dwarf cherries).

Note: PERA triads have an evolving relationship with adjacent ANA triads. In most cases the PERA triad will slowly gain influence and change adjacent ANA triads through shading, soil character adjustments, etc. We can use this to our benefit by customizing our crops to that niche: shaded summer spinach, protected overwintered onions and eventually woodland herbs.

It is also important to see the evolving PERA triads need for the space of adjacent annual beds for management access to the PERA triad for harvest, pruning and other management activities. There are also many opportunities for use of PERA triad products on adjacent ANA beds.

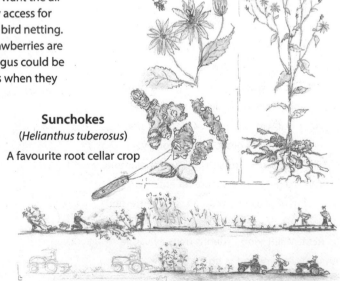

Sunchokes

(*Helianthus tuberosus*)
A favourite root cellar crop

Technique should be flexible like pencil, erased and redrawn as needed

ANA Triad Design Considerations

There are many factors to consider when redesigning ANA triads, including these primary considerations.

- **Umbrella management:** ANA triads should be primarily focused around similar management needs. Put these crops under the same umbrella to improve efficiencies. For instance, crops in a plot which are all irrigated with drip tape (mulched crops) or sprinklers. Or crops in a given plot all require hilling (row crops).

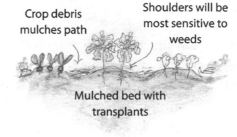

Crop debris mulches path

Shoulders will be most sensitive to weeds

Rolled/crimped cover crop becomes mulch for crops planted into it.

Mulched bed with transplants

Umbrella Management of ANA Triads

Understanding crop production needs can help assemble crops into triads for easy umbrella management.

Row crops benefiting from hilling, and few rows per bed are easily managed for weeds.

Umbrella Management
Crops with similar management needs can be grouped together. This is a good start for building crop guilds on a market garden.

Early mixed vegetables have small seeds and row cover is useful for insect protection and extra spring boost.

Fine seeds

Crops with fine seeds like turnips, radish and carrots require a different bed preparation than larger seeded crops. They are more vulnerable to poor soil-to-seed contact, erosion and overexposure to the elements.

Hilling

Some crops enjoy hilling to ensure crop quality and plant support. They also lend themselves to easy row-style management (discs and hoes) because they occupy only one or two rows per bed.

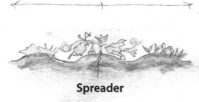

Spreader

Some crops will occupy a considerably more space later in the season. Planning for this expansion is critical to avoid overtaking unharvested crops. Design for cover crop integration and maximize use of adjacent beds before sprawl.

GREEN THUMB TECHNIQUE

Path mulching

Woody material from raspberry canes and tree pruning can be chipped into adjacent paths for mulch and balanced with dwarf white clover for path weed management and in situ composting.

- **Guild crop production:** Triads are organized for agro-ecological services and companionship. These considerations combined with umbrella management factor to determine crop guild triads.
  - PEAS/POTA/BEAN → Row crop guild
  - CARROT/SALAD/CARROT → Early mixed vegetable crop guild

ANA Triads

Umbrella management and guild crop production are organized together

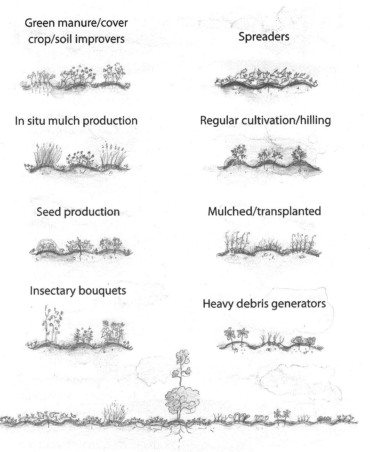

Green manure/cover crop/soil improvers

Spreaders

In situ mulch production

Regular cultivation/hilling

Seed production

Mulched/transplanted

Insectary bouquets

Heavy debris generators

Ana triads can fit between PERA triads and are rotated through time and space.

Crop guild triads can partition space and resources, improve pest management and provide other beneficial interactions within the triad and for the greater garden ecology.

- **Alternate maturity patterning:** Every bed should alternate maturity across the whole ANA plot.

Days to Maturity

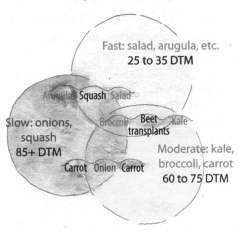

Fast: salad, arugula, etc. 25 to 35 DTM

Arugula Squash Salad

Slow: onions, squash 85+ DTM

Broccoli Beet transplants Kale

Moderate: kale, broccoli, carrot 60 to 75 DTM

Carrot Onion Carrot

- **Guild unit triads:** This entails variations of crop guild triads to help manage alternate maturity patterning across the whole permaplot and flexibility in desired quantities of key and neighbor crops.

  An example of early mixed vegetables GUT for May 1:
  - Outer bed: Salad
  - Center bed: Carrot
  - Outer bed: Spinach

    Succession in July may look like this:
  - Outer bed: salad/spinach
  - Middle Bed: carrot
  - Outer bed: dill/cilantro
- These are both guild unit triads of the same crop guild triad

Guild Unit Triads from our Guild Crop Rotation Demonstrate Alternate Maturity Patterning

(see Days to Maturity Diagram for relative DTM of bed by color)

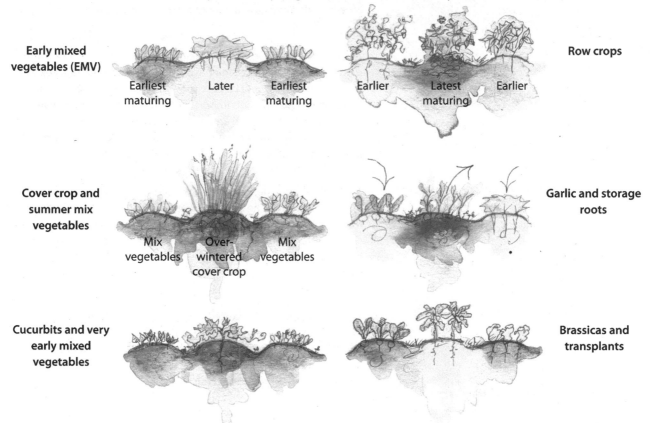

Early mixed vegetables (EMV)

Earliest maturing · Later · Earliest maturing

Earlier · Latest maturing · Earlier · **Row crops**

Cover crop and summer mix vegetables

Mix vegetables · Over-wintered cover crop · Mix vegetables

Garlic and storage roots

Cucurbits and very early mixed vegetables

Brassicas and transplants

- **Guild unit plot patterning:** Understand how different guild unit triads (GUTs) can assemble together to maintain alternate maturity patterning across a permaplot and flexibility for seasonal production constraints like market demand and weather.

Mixing Guild Unit Triads Using Alternate Maturity Patterning

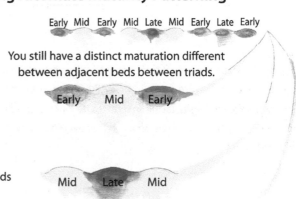

Early Mid Early Mid Late Mid Early Late Early

You still have a distinct maturation different between adjacent beds between triads.

Early · Mid · Early

Different crop guilds can be placed together in a permaplot without jeopardizing alternate maturity patterning.

Mid · Late · Mid

Early · Latest · Early

Alternate Maturity Patterning Examples

✓ **This is good**

Squash & EMV

This is good because crops partition resources and space well. Fast growing salad greens make room for slow-growing squash.

✗ **Don't do this, you need access sooner**

Broccoli and onions

This could be improved despite being a mix of slower maturing and moderate maturity crops. You have to wait 2 months for broccoli to be harvested before you can use grow and mow management. In a mulched triad like this it is important to have early access to manage the edges between mulch and path. You are better off alternating the onions with transplanted beets. Also, if your row cover ever gets loose from the broccoli, it will bend over all your onions!!!

☺ **Okay, but too similar in maturity**

Storage roots

This is okay. But it depends on the timing. If these are both seeded for early crops then the timing is pretty much the same. However, if the beets are transplanted (40DTM) and the carrots are the main crop (75DTM), that is manageable. But I would prefer to put lettuce between the carrots.

☺ **Right species, wrong order**

Potatoes, beans, peas

The potatoes, unless harvested very young, will be in the ground much later than beans or peas. They should be placed in the middle. Use this as an integrated fallow triad where cultivation is used intensely between rows. This whole triad can be cover cropped into oats or rye with a broadcast seeder later.

GREEN THUMB TECHNIQUE

We plant peas, beans and potatoes together because they all need to be hilled. We use a small 11 hp Allis Chalmers G tractor for this job. We can easily tackle several triads for fast, effective management. Then the whole area can be covered cropped with a broadcast seeder.

ANA Triad Examples

Notice how these examples include umbrella management and crop services

1. Center crops spread over early-harvest neighbors (salad/squash/baby spinach)
2. Both are pest confusers (storage carrots/garlic/storage carrots)
3. Neighbors confuse pests (radish/squash/arugula) ➜ residual radish left to flower for habitat
4. Center crop cools neighbors (summer salad/rye cover crop/summer spinach)
5. Neighbors are harvested early for central access (EMV/zucchini/EMV)

6. Entire triad is all bed transplanted (beet/ broccoli/head lettuce) for umbrella management

7. All beds hilled (peas/potatoes/beans) for umbrella management.

Crop-cover Crops

One of the reasons we like to use alternate patterning is for the benefits of crop-cover crops.

Crop-cover crop is what we call any typical garden crop that can be left to produce a cover crop. These generally are leafy vegetables whose plant, after harvest, is still in the ground and capable of growing. They can provide much-needed benefits like shade, windbreak, habitat and organic matter without the time/space/energy needed to prepare the bed and seed cover crop.

- Arugula, for example, can continue growing after being cut for salad, as opposed to a carrot that is pulled up for a bunch.

- Or consider leaf lettuce as opposed to a head lettuce (cut at the base). For this reason we grow way more leaf lettuce than head lettuce.

Other crop-cover crops we use often:

- Spinach for cover crop
- Radish for habitat, pest trap crop and green manure
- Mustard greens for habitat and anti-nematode
- Overwintered kale for pest trap crop and improving soil structure

GREEN THUMB TECHNIQUE

Cover Crop Patterning

Integrating cover crops with your crops will increase their productivity. Alternate maturity patterning maximizes this.

- Cover crops scavenge excess nutrients from compost applications.
- Cover crops receive sprinkler irrigation when it's dry.
- Cover crops have a longer growing window as they are seeded as beds become available, no waiting for an entire field to be ready.
- Cover crops receive a mulch down from mowed crop debris and weeds of early crops.

Radish

After radish harvest is done there are always over-mature and split radish that can be left. These will flower, attracting beneficials to the nectar and feed on flea beetles. Spiders will build webs in the lofty stalks. Also, young radish seeds are edible.

Lettuce

After several harvests, lettuce will produce tremendous towers of succulent leaves. It is very competitive, nitrogen-rich and a good insectary.

Beans

When left to mature, beans can return many nutrients and materials into the soil. We harvest beans for peak market demand then leave them for dry beans and/or cover crop.

Any of these, if left in the garden can become a cover crop, producing towering organic material (lettuce) or continuing to mine the soil (kales).

Fitting Cover Crops Intensively into Your Rotation

You finish harvesting your lettuce, you leave it to grow and tower and flower. Benefits include habitat and food sources for beneficial insects, nutrient accumulation from soil, organic matter production and garden cooling. You save the cost of cover crop seed and land preparation. I see this as the critical way of maximizing cover crops in season when farms are usually strapped for time and space.

Some crop-cover crops are more easily incorporated for immediate seeding/planting and even used as In situ mulch is mulch in which you grow a crop-cover crop and then flail mow it. Next, seed or plant crops into the debris.

Specialty Triads

We can easily designate a triad for a special purpose. This can include:

- Caterpillar tunnel
- Hoop house
- Buffer (along a grass road)
- Hugelkultur compost beds
- Chicken tractor or egg-mobile
- Other livestock

For instance, we make specialty triads for our husk cherries. These triads are reformed several times for increased height. Only the tip-top of the reformed beds is harrowed for planting. Than we cover it with wide piece of weed barrier and plant into premade holes. This allows the husk cherries to ripen and fall down into a clean weed barrier trough where we can scoop them up.

GREEN THUMB TECHNIQUE

Lettuce as a Crop-cover Crop and In situ Mulch

- Seed, maintain and harvest several successions of lettuce
- Let the lettuce bolt, flower and produce incredible nutrient-rich organic matter.r
- Flail mow 2 weeks before needed for crop
- Apply black tarp and fry the bed down (repeat again for false sowing
- Transplant fall broccoli and kale

GREEN THUMB TECHNIQUE

Additive Planting

Additive planting, a form of guild bed patterning, is a method of adding beneficial plants into your transplant mix at a certain ratio in order to achieve desired benefits.

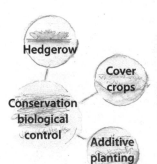

Alyssum/lettuce density research indicates that fewer allysum are needed than originally thought, making the cost/benefit quite favorable.

The results are further enhanced when combined in a three-pronged conservation biological control model for IPM.

Guild Crop Production

The Natural Companions

Nature is a wild place where the fittest survive. This leads to an ecosystem with many species in their own niche, specialists and generalists. My local forest has shade-tolerant and sun-loving trees, water lovers by the streams and dry survivors in uplands. Some flower and fruit in spring and others in fall. And this is just the trees. Add in other diverse perennial layers, fungi, bacteria, arthropods, mammals, birds and amphibians, etc. We have a functional community with many interactions between organisms. This involves competition, but also organisms are supporting and balancing each other.

Symbiosis

According to my old biology textbook, symbiosis is an ecological relationship between organisms of different species that are in direct contact. This may involve a larger *host* and smaller *symbiont*.

Companion Planting

In garden terms, a companion crop is a plant that benefits or is benefitting another.

Symbiosis can include:

- Mutualism (where both species benefit), for example, mycorrhizal fungi and squash.

Well-planned companion planting takes advantage of symbiosis and umbrella management.

The fungi help squash access nutrients and water, and the squash provide sugars.

- Commensalism (where one species benefits, but the other is unharmed), for example, hardy kiwi growing up a fruit tree.

The kiwi has increased access to light and the fruit tree isn't bothered.

Symbiosis is from ancient Greek; syn (together), and bioisis (life)

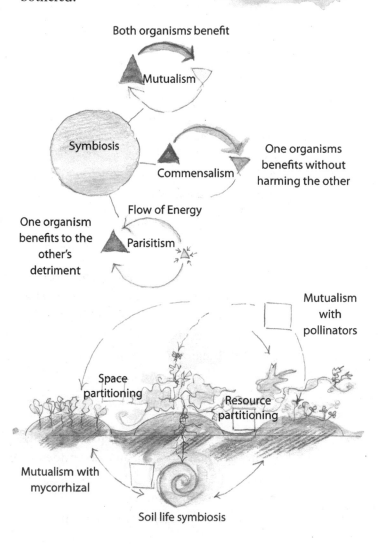

Both organisms benefit

Mutualism

Symbiosis

Commensalism

One organisms benefits without harming the other

Flow of Energy

One organism benefits to the other's detriment

Parisitism

Mutualism with pollinators

Space partitioning

Resource partitioning

Mutualism with mycorrhizal

Soil life symbiosis

- Parasitism (where one species is harmed as the other benefits), for example, chaga, which grows from yellow birches in our woods.

When we talk about companionship, it is often very vague. These plants are good neighbors, or these plants work well together with maybe some reference to pest deterrents or space sharing. I think it is important to define exactly what plants do to help each other so we can more easily design with them to form crop guilds. This is especially important in market gardens where you are putting in long beds and large plots of species, which should companion, but must also be efficient for field operations!

The entire garden is a stage! I think of the garden as a stage and the plants and other species as actors performing roles. Some are chameleons and can act in many roles, serving many needs. Others are very specific, type-casted to a particular role. Favorites arise as stars of the garden and are used in many locations to perform throughout the years.

The Guild

A guild is a grouping of organisms, working well together through direct and indirect aid of individual or mutual productivity.

Guilds can include annuals, perennials, animals, fungi, soil bacteria, humans, etc. Examples include:

- ANA guild triads,
- PERA guild triads,
- Mixed guilds, including perennial guilds that we integrate with annuals in their early years.
- Guilds integrate animals
  - Rotating chickens into your gardens

- Guilds integrate soil life
  - Enhancing soil life to interact with the garden
- Guilds integrate humans
  - You are part of the ecosystem

Guilded Community

Services worth more than gold

Air · Wildlife · Perennials · Big animals · Cover crop · Farmer · Crop · Water · Neighbor crop · Soil life · Small animal · Mineral

Soil food web and its myriad interactions are the most powerful guild services.

The Simplest Guild: Bed, Path and Soil

A guild unit is the organizational unit for designing guilds. The most useful organizational unit for market gardeners is the guild unit triad. The most basic guild unit, however, is the permabed: with bed crop, cover-cropped path and the soil life beneath partitioning resources and mutually supporting productivity.

Simplest Guild

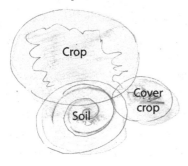

Soil Life and Cover Crop Relationships

- Active mycorrhizal fungi exchanging soil nutrients for plant sugar in a mutual symbiosis
- Soil bacteria and clover cover crops fixing nitrogen that becomes available to the crops
- The crop and clover returning food and improving habitat for soil life

That is right, simply building healthy soil moves you toward guild crop production! All those organisms are waiting to make relationships with your plants.

Guild Companionship

Guild companionship is the selection and routine assembly for production of specific species with the goal of umbrella management and symbiotic possibility. In the permabed system we design crop guild triads where species occupy beds within a triad to form relationships between key crops, neighbor crops, the soil life and other chosen organisms. There are many different ways that crops can form companionships and this can be helpful in designing crop guilds. David Jacke, in his seminal work the *Edible Forest Garden*, makes good discussion of three broad types of guilds. Note: crop guilds may belong to more than one category.

Resource Partitioning

Crops may require the same resources: sunlight, water and nutrients, but they use them at different times of day, season or from different layers of the canopy and sub-surface. This difference in access from variable root type/depth as well as leaf shape and canopy form/height can greatly enhance the capacity of a piece of land in production. No sunlight or water is wasted. This will also reduce weeds, which tend to fill vacant niches in disturbance agriculture.

In ANA triads we partition resources through time and space. For instance, in our cucurbit guild, the lettuce beds are harvested by the time the squash is rampant. They do not compete, despite the squash using the entire space at one point in time. Furthermore, the lettuce helps mulch out weeds that may have become an issue in the squash production, avoiding periods of bare soil.

Cucurbit Guild

On the other hand, in PERA triads, resource partitioning is achieved through mixing species to harmonize their form and function. Take this bed patterning: maple, gooseberry, pear, Siberian pea shrub, oak, with a ground cover of thyme and paths in white clover. This bed will share resources much better than a row of just one species. These species have different forms above and below ground and each has its own unique lifecycle. Different nutrients are in demand by different species. In general, crops require certain nutrients for growth, flowering and fruiting. If the species in an area flower and fruit at different times, they will partition nutrients better.

Mutual Support

These occur when one or more species' natural character, habit and products meet the needs of another species. The most important example is the beneficial relationship between mycorrhizal fungi and your crops. These exchange sugars, nutrients and other elemental flows with each other. Another example would be the use of oats as a trellis for peas. The peas can climb, and the oats get nitrogen.

Winter-killed Cover Crops

Oats & peas
(*Avena sativa &
Pisum sativum*)

Winter-killed cover crop combination. Grows well in cool season, fixing nitrogen and building good root organic matter. Oats tolerate a wide pH range and enjoy moist soils. Pea flowers attract bees. Pea root exudates interfere with germination of lettuce seed. There is great biomass production that is easily incorporated.

Oilseed radish
(*Rasphanus sativus*)

Very good at loosening the soil structure and creating biopores. A succulent nitrogen scavenger, it easily releases the nitrogen come spring as it quickly decomposes. Some varieties have glucosinolate levels high enough to suppress nematodes.

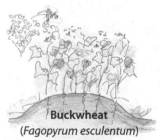

Buckwheat
(*Fagopyrum esculentum*)

Accumulates insoluble phosphorous (phosphate) and potassium while shading out summer weeds. Attracts flea beetles and shouldn't precede brassicas. Overall organic matter contribution is low.

Community Function

In this case, species are helping each other without a direct interaction but rather contribute to an overall functioning community.

For example, early flowering cover crops in adjacent beds provide nectar for pollinators of fruit trees and/or habitat for predatory insects that prey on pest insects of a neighboring crop. The flowering cover crop is not directly succoring his neighbor (the fruit tree). But there is a community benefit!

 Sometime parasitism is a good thing. In nature death brings life. Parasitism in the soil food web helps free up nutrients for crop plants from decaying soil microorganisms. We certainly enjoy when parasitoid wasps inject their progeny into pests, like leek moth larva, and bring this pest population down.

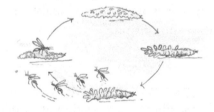

Companionship Examples
Cover Crops and Guilds

Cover crops are a primary crop guild component, because they can serve guilds in season through alternate patterning and because they are essential community supporters within a crop rotation. Clover cover crops provide nitrogen to future crops, buckwheat suppresses weeds and rye holds the soil and protects it over the winter and in spring.

Integrated pest management means you use many strategies to make sure your crops are pest free. Companionship in crop guilds can be of use in IPM.

- *Pest confusers* are strong-smelling plants that confuse pests.
  - ○ Mint is a great one, especially when mowing releases its oils.
- *Pest deterrents* are plants that deter pests.
  - ○ Thyme repels cabbage loopers
  - ○ Chives repel carrot rust fly
  - ○ Nasturtium repel many insects, including squash bugs
- *Pest trap crops* attract pests and hold their attention.
  - ○ Eggplant is an excellent trap crop for Colorado potato beetle. The best way to trap pests of a crop is to plant a row of the crop near last year's triad. We always plant a row of potatoes near last year's potatoes and rotate this year's crop very far. Only 50% of potato bugs will make a journey over 2,000 feet. In small plots, a handful of eggplants will trap your potato bugs so you can deal with them efficiently.
- Beneficial and predatory insect habitat
  - ○ Lots of flowers through the season help
  - ○ Mixed flowering and treed hedges.

Do You Know?

Niche
A crop's niche is where it fits into the Earth's ecosystems. By understanding a crop comprehensively we can best find its niche within the market garden and companion it appropriately into crop guilds. Its niche includes its form, function and interactions within a community. This is why monocultures are inefficient; the whole field is filling the same niche and so they are all competing with each other.

I record the flowering dates on our farm and try to make sure there is always something flowering!

Efficient Guilds
Most modern crop production is monoculture, guaranteeing an entire field uses the same resources (nutrients, water, space and light) at the same time and in the same way.

The most intensified and diversified annual guild would be a mix of six or more species seeded into a bed. This can work! However, diversity also entails increased logistics and potential inefficiencies or failures may arise. As permaculture market gardeners we must prioritize efficiency of space/time/energy, which means maximizing ecosystem services without jeopardizing market garden efficiencies. Therefore, crop guilds should strive for umbrella management as well as companionship. Any major loss in efficiency due to a crop guild arrangement must be matched with significant benefit from the new relationships.

We do intercrop within the bed; rows of carrots and radish and transplanted lettuces with tomatoes, etc. But our standard is the triad with three crops and cover crops. From here can diversify to multiple crops per bed, but the basic crop guild is three crops per triad with cover-cropped paths.

Remember: The two most time consuming field jobs on a market garden are weeding and harvest. So consider these foremost when designing crop guilds. It doesn't make sense to build a guild that interferes with each other's harvest.

Garden Community Relationships

Many relationships exist, including:

- Crops within a bed in different rows
- Crops within a bed in blocks
- A bed's crop and the cover-cropped paths
- A center bed with any neighboring beds in a triad
- A triad and relationships with adjacent triads
- Any past or future crop guilds in the rotation and their relationship with current crop guild
- Adjacent PERA guilds
- A crop and the permaplot or all permaplots

Do You Know?

Great crop service opportunities

- Integrate and benefit from cover crops in our paths
- Maximize benefits from neighboring beds (confuser, shade, in situ mulch production)
- Use field roads and crossbeds to maintain permanent diverse ground cover alleys
- Alternate the species and maturity of all beds through the garden blocks
- As opposed to field agriculture, permanent beds allow you to plant whatever you want wherever you want and manage it, as it requires.

Guild Unit Triads (GUT)

Remember: You can build a few variations of any one crop guild. These variations help balance your desired quantities of production for each crop and deal with seasonal production restrictions. They also allow for special purpose designs and placement within the garden. Maybe a GUT with specific trap crops is needed in a given plot to improve pest management there.

Preceding greeting, proceeding feeding.

THE SCOPE AND SCALE OF GUILD RELATIONSHIPS

Services Through Time and Space

1. **Preliminary and fallow services:** This includes use of cover crops in planned land preparation and fallow periods, such as clovers, rye and oilseed radish and/or the use of livestock, to improve soil structure, fertility and organic matter before garden crops are ever grown.

2. **Intra-permabed services:** This includes intercropping within the same bed: row by row for the whole length and/or section by section along the bed's length. Also, consider the interaction between beds and their paths: nitrogen fixing, sharing space through time, etc.

3. **Intra-triad services:** This involves direct relationships between crops in the same triad, such as provision of mulch material and shading against extreme heat, and services between key crop(s) and neighbor(s).

4. **Inter-triad services:** This entails relationships between adjacent triads in a permaplot.

5. **Seasonal rotation services:** This pertains to the services to the following crop in the rotation; including taprooted crops loosening soil or aggressive crops outcompeting weeds for less competitive crops. Remember the mnemonic: Preceding greeting, proceeding feeding.

6. **Intra-permaplot services:** This includes services to other crops within the plot, such as those beneficial relationships to be had with perennial triads and the rotated annuals in between. This also includes crops acting as trap crops against pests or as deterrents to confuse pests.

7. **Inter-permaplot services:** This could entail nectar sources and habitat for beneficial organisms and the overall cooling and microclimate effects of trees, shrubs and some annuals. Many relationships are found between plots in a larger field, just as there are relationships and benefits to be had between the lake shorelines and the species that exist within its depths, surface and adjacent lands.

8. **Seasonal rotation and future garden services:** This includes all crops which have a positive impact on the overall soil and garden ecosystem health that will reverberate far into the future, especially as it allows movement toward a resilient woodland ecosystem garden.

What Is a Guild?

In our context, a guild consists of organisms in community, many of them benefiting directly or indirectly from others.

Layers within the guild will increase the above- and below-ground diversity and potential resource partitioning.

Scales of Relationships

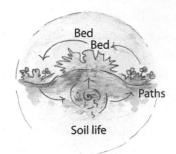

Intra-permabed services
The bed crop, path cover crop and soil life form the most basic relationship and potential guild. Must foster soil health to maximize this most essential relationship.

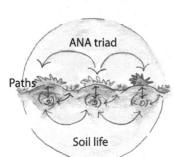

Intra-triad services
Companionship between crops, cover crops and soil within a triad is the next level of organized relationships. This is the basis of a crop guild.

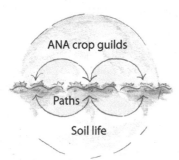

Inter-triad services
Relationships between ANA crop guilds within a plot.

9. **Whole Farm Services:** At this level we add the crucial element of human relationship with the garden. We benefit from the harvestable yields and other agro-ecosystem services to the whole farm. This is the level where the garden area interacts with other projects and processes of the farm for their mutual benefit.

10. **Greater Community Services:** At this scale we see a farm community guild, where farms form interrelationships. Each has their own guild enterprise production and the farms begin to *share tools, inform decisions, cycle waste and, overall, balance each other's productions.*

11. **Regional Services :** Might include water purification, habitat provisioning, foodshed security

12. **Global Services:** At this scale focus should be on knowledge sharing and carbon sequestration.

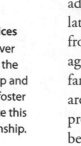

The shorter the chain between raw food and fork, the fresher it is and the more transparent the system is.
— Joel Salatin, brainyquote.com

Principle

Three-scale Management

In order to better manage the complexity of crop guilds we operate on three scales:

• Bed scale • Triad scale • Permaplot scale

Each scale has specified operations that work well to achieve best management practices and field operation efficiency.

Bed scale:

• All field equipment is operated at this scale: reforming, harrowing, mowing, seeding. We make a pass with equipment up one bed.

Triad scale:

• Most row cover, black tarp culture, and cover cropping.

Permaplot scale:

• Crop rotation (for farms of more than a few acres),

• PERA triad selection, management and evolution.

Intra-permaplot Services

Relationships are formed between ANA crop guilds in rotations with the anchored PERA guild.

Seasonal Rotation Services

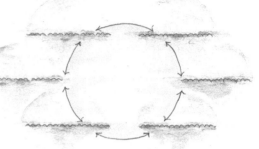

Inter-permaplot Services

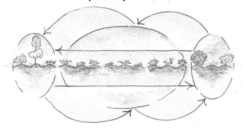

ANA triads form relationships with PERA triads and PERA triads form relationships with each other.

Crop services are provided and received through the seasons and in the long run for the whole garden ecosystem.

Whole Farm Services

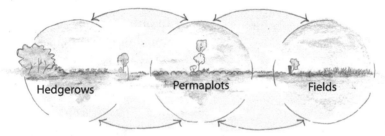

Hedgerows   Permaplots   Fields

Relationships exist between various enterprises and the farmers and other ecologies.

Farm community guild: Can farms begin to balance each other better by broadening the marketplace with complimentary productions?

Greater Community Services

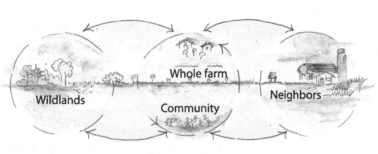

Wildlands   Whole farm   Neighbors
Community

Regional and Global Services

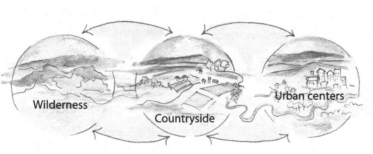

Wilderness   Countryside   Urban centers

Regionally, goods and services are exchanged, but globally relationships should focus on education, policy and galvanizing agro-ecological team spirit.

Guild Crop Designs

HOW TO GET STARTED

We are always striving to build crop triads into guild crop triads. The first focus is always on the soil because this is the start of any good guild. Healthy soil life, as we have discussed, is a fundamental principle of permaculture farming, without which any further complication of guild patterning is really unneeded. The next step is creating truly important relationship benefits between crops within a triad. A routine of diverse planting is a good idea, but eventually you can intentionally intercrop for specific known benefits.

Simplest Guild

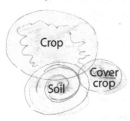

Guild Crop Production Timeline

- Build permabeds and improve soil with sandwich bedding
- Organize your crops into triads with umbrella management
- Integrate perennials into your garden for enhanced diversity
- Build a crop rotation around building soil and enhancing soil life

- Design production triads into crop guilds

Guild Companionship Design

Design crop guilds for services in this order, unless you have specific needs. This is what I think is most useful for commercial growers:

- Soil health/fertility/weed suppression
- Resource/space partitioning
- Deterrents and confusers
- Predatory insect habitat, etc.

In order to discover crop relationships for potential umbrella management and symbiosis, we must first conduct a critical crop analysis.

Critical Crop Analysis

This means study the crop.

- When does it mature? → Research the DTM on the seed packet
- What does the crop's root look like? → Dig a plant up when it is mature
- What is the species' full life cycle? → When you harvest a crop, just leave one in and watch it for a full year.
- Let it flower, mature, fruit, die, decompose. What does the ground under it look like in the spring?

Guild Design Management

We need to design our crop guilds as living examples. Living design is key because it explores real-time scenarios for us to observe and refine the crop choices, production strategies and guild services. Design management makes use of research beds to avoid obstructing seasonal field operations and maximizes research visibility. Once a new guild design has been trialed and redesigned, then it can be put into full production. We can adopt, adapt and evolve guild designs from other farms starting with the guild crop builder, or use it to start from scratch.

Guild Crop Builder

Crop Guild Name: Pick a key crop or express a guild idea.

Objectives: Why are you designing this guild? To make use of a certain garden soil, meet market demand for early spinach, improve pest management of potatoes, etc.?

Constraints: List constraints that must be considered for efficient & effective production, given the tentative scenario created.

Guild Unit Triads:
Key crop(s) can either be in yellow or red. Assess your demand.
Crops in higher demand for a given planting should use outer beds.
Organize DTM: If red equals fast maturation, than yellow equals slower or vice versa.

Guild Crop Triad

Consider possible benefits through time

	Key crop	Neighbor	Key crop	**Path crop**
Last year:				
This Year:	Neighbor	Key crop	Neighbor	Consider seasonal rotation services:
Next year:	Key crop	Neighbor	Key crop	
DTM				Purple column for path crops
Debris				
Habit				
Edible parts				
IPM				
Water Management				
Fertility				
Harvest				
Inoculants				
Benefits from				
Benefits provided to				

Analyze potential Services within triad and for future guild crop triads

GUILD CROP PRODUCTION PRINCIPLES

Build Relationships around a Key Crop

When considering crops for a triad, focus on improving the production of the key crop first, then consider the neighbors. Focus leads to further solutions!

Triads should never host problems for proceeding crops.
Grasses and especially pasture host wire worm, and soybeans host rizoctonia.

Triads should improve conditions for proceeding crops.
Or on a broader scale, a crop could improve the soil for crops one or more seasons later, such as winter rye stabilizing soil through the winter for summer crops.

Bed management for proceeding crops should not negate benefits left by preceding crops.
For instance, many brassicas like kale leave great root systems in the ground that can decompose to form awesome biopores. If we pull the stalks and roots out or plow them under, we remove or alter this benefit.

Design to Use Crop Services

When farmers know and value a crop service then they design to maximize it and avoid systems that destroy it. Designs may include modifying crop rotation, equipment and land preparation techniques.

Plant for the greater improvement of the garden ecosystem:
For instance planting a row of oak trees helps to cool your entire garden ecosystem.

Scale of Relationships and Services

Relationships between organisms occur at all scales. Benefits from crops can be felt within a triad and across an entire permaplot or throughout the farm. It is important to begin to understand the density of a crop required for its needed benefit and how various ecologies on the farm contribute to overall beneficial interactions.

- For instance, early flowering meadowlands attract pollinators until fruit flowers are out.
- One row of maturing trees can serve as a windbreak for a whole field.
- A bed in mint can help confuse pests for an entire permaplot.

Integrate bed elements for guild production

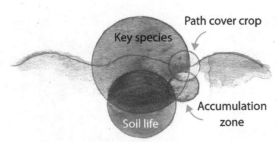

Consider soil life interactions and conservation

Integrate triad elements for efficiency, services & bed reforming

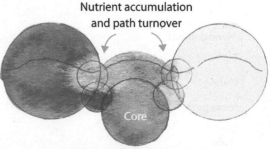

Soil life recolonization from below

Consider preceding and proceeding production and integrate into guild crop rotation

Guild Crop Production Patterns out Across the Permaplots Through the Year

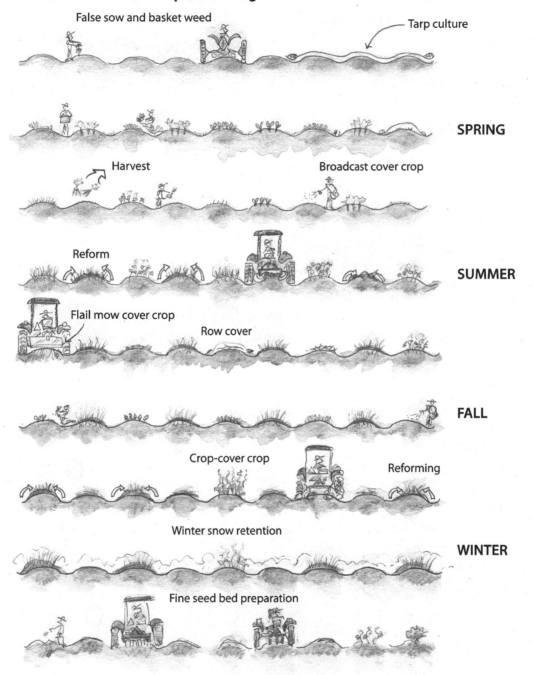

False sow and basket weed

Tarp culture

SPRING

Harvest

Broadcast cover crop

Reform

SUMMER

Flail mow cover crop

Row cover

FALL

Crop-cover crop

Reforming

Winter snow retention

WINTER

Fine seed bed preparation

Do You Know?

Biopores feed soil life, which line the new void space with available nutrients. They also vastly improve soil structure for the movement of air, water and nutrients and crop root growth.

Essential Guild Services to Include in your Garden Plan:

- Food production, or other marketable products
- Nitrogen fixer (clover in paths, nitrogen-fixing trees in PERA triads)
- Nutrient accumulator (dandelion and plantain in paths) (summer green manure crops like buckwheat and red clover)
- Pest deterrent (aromatic plants in paths or crossbeds)
- Beneficial habitat (let crop go to flower, path plants)

- Soil stabilizer, protector (winter cover crop)
- Soil structure improvers, oilseed radish, kale (leaves biopores)

These can be grown in/as:
- Crop on bed top
- Green manure
- Cover crop
- In paths
- In crossbeds
- In laneways
- In hedgerow
- In riparian buffers
- In windbreaks

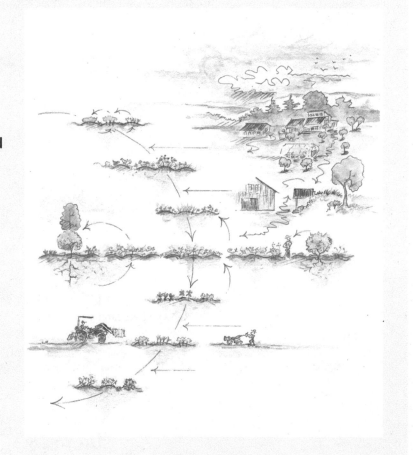

Principle

Full Cycle, Multifunctional and Integrated Design

Our permabed system is designed for operational rhythms, considers crop and ecosystem services and works to integrate productions within crop guild, and garden patterning.

Guild Crop Rotation

Once we design guild crop triads, and their GUT variations, we need to put them into motion through time and space. This rotation allows us to have all the benefits of the crop's services spread through the garden ecosystem. Remember design management — the process of working with your designs to improve them through trialing, observing and redesigning. Now is the time to being to trial your guild crop triads in motion through time and space. This diagram below can help conceptualize this.

Simplified Guild Crop Rotation

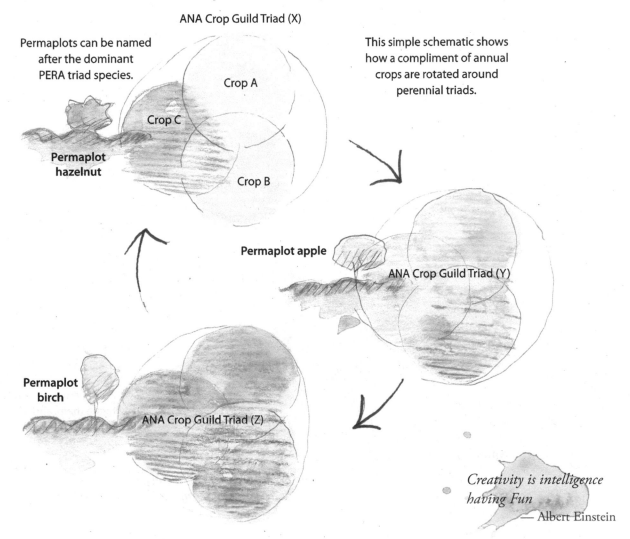

ANA Crop Guild Triad (X)

Permaplots can be named after the dominant PERA triad species.

Crop A

Crop C

Permaplot hazelnut

Crop B

This simple schematic shows how a compliment of annual crops are rotated around perennial triads.

Permaplot apple

ANA Crop Guild Triad (Y)

Permaplot birch

ANA Crop Guild Triad (Z)

Creativity is intelligence having Fun
— Albert Einstein

Guild Crop Rotation

Crop guilds have multiple services through space and time

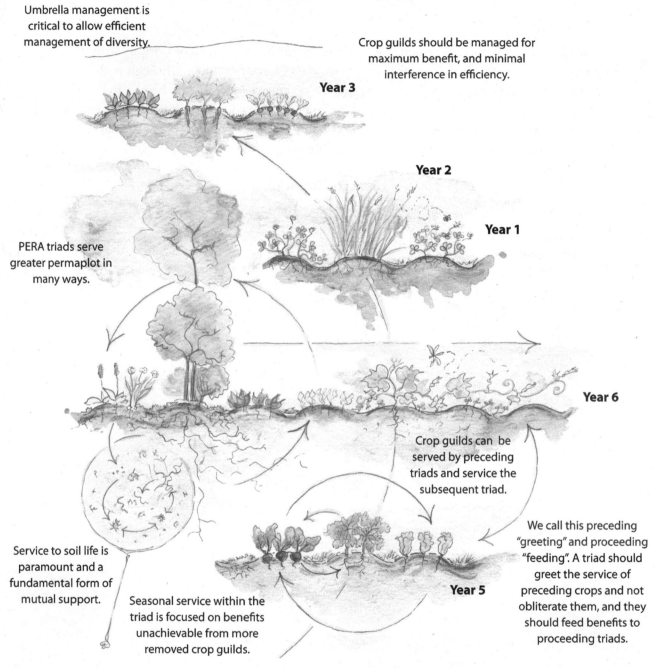

Umbrella management is critical to allow efficient management of diversity.

Crop guilds should be managed for maximum benefit, and minimal interference in efficiency.

Year 3

Year 2

Year 1

PERA triads serve greater permaplot in many ways.

Year 6

Crop guilds can be served by preceding triads and service the subsequent triad.

We call this preceding "greeting" and proceeding "feeding". A triad should greet the service of preceding crops and not obliterate them, and they should feed benefits to proceeding triads.

Year 5

Service to soil life is paramount and a fundamental form of mutual support.

Seasonal service within the triad is focused on benefits unachievable from more removed crop guilds.

Crop guilds should never host problems for guilds that follow!

GUILD CROP TRIAD SUCCESSION

Part of your guild crop rotation is the sucession of triads within the same season. Understanding how seeded beds can be turned over or mulched beds can be replanted. Basically;

- Maintain alternate maturity patterning
- Take advantage of available niches quickly through relay cropping or popping in transplants soon after a bed is spent.

Grow and Mow Management:

An important part of your guild crop rotation is integrating cover crops and grow and mow weed management. This includes:

- Growing overwinter cover crops and mowing them at the ideal time to enhance bed health before cropping
- Allowing crops to become cover crops and mowing these when they mature

Early Mixed Veg Triad, GUT variations

These are variations of the same crop guilds, similar species maintain alternate maturity

Spin Carrots Salad

Carrots Spin Carrots

Carrots Salad Beets

Dill Carrots Cilantro

Transplant Triad Succession

Onion Beets Onion

Harvest beets and replant into the mulch with kale and lettuce

Lettuce is harvested, broccoli matures

Onions are harvested and replaced with transplanted fall spinach

Next year

Broccoli left standing as snow fence to help spinach over-winter

Spinach matures early

Debris is eventually mowed and bed is prepared for summer squash

DTM color key:

- Blue: Longest maturing
- Red: Moderate DTM
- Yellow: Fastest maturing

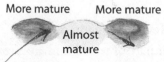

Triads have their own succession of maturing crops

More mature

More mature

Almost mature

(Center bed is harvested)

Next

First to mature

Next

Almost mature

New crop will be the THIRD

Almost mature

- Overseeding early maturing beds with cover crop seed and mowing the crop debris and any annual weeds over top as mulch
- Growing path cover crops to add to the guild and balancing these through time and space so they work well with your rotation

Design Management for Weeds

This is an example of a full-cycle approach to weed management we have evolved using observation, trial and redesign

Weed-free compost is used in season

Pull emergent weeds to avoid "weed bomb" next spring!

Stale seed and kill emerging weeds with rakes, basket weeders or flame weeders

Alternate the maturity of crops in a triad to allow timely access between later maturing crops for easy weed management with mowers of all sizes

Cover-crop paths to fill the niche weeds would usually occupy

Flail mow crop debris over top of seeded cover crops for improved germination of broadcasted seed (no till)

Use micro sprinklers, timely seeding and row cover to enhance your crop's competitive edge against further emerging weeds

A guild crop plan is like a crop plan, but you organize and balance your production into guilds.

Manage your triads as per guild crop plan

- For instance, don't seed white clover down in the paths of a bed in at the end of April only to turn the bed over in August
- Taking advantage of an early crop's spent bed as an access point for tractor mowing of the bed, and offset mowing of the cover cropped paths

Our Current Guild Crop Rotation

Our rotation is based on umbrella management and crop services. It considers land preparation, cover cropping, harvest efficiency and tool/supply sharing. They also serve to provide complementary diversity for IPM, adhere to alternate maturity patterning for access, cover crop priority and grow and mow management. It emphasizes space/time partitioning, flowering diversity and in situ compost production. It is evolving.

1. **Cover Crop/Summer Mixed Brassicas (EMB) (1st year)**

a. Management specifics: Last year's storage root (carrot, beet) beds were cover-cropped to rye. Now they can be mowed and power-harrowed for summer mixed brassicas (SMB) for June-July seeding.

b. Beds that were in garlic are now in red clover and rye, rye is mowed, and red clover takes over to help balance the high carbon mulch with nitrogen.

c. Beds in rye are prepared for summer mixed brassicas and row cover is used for flea beetle protection.

d. GUTS: red clover/rye/red clover ➝ red clover/radish/red clover and arugula/red clover/salad turnip

e. Red clover is tall and helps shade and cool summer mixed brassicas for optimal quality. Also, provides nectar and habitat.

This is the guild crop rotation we began using

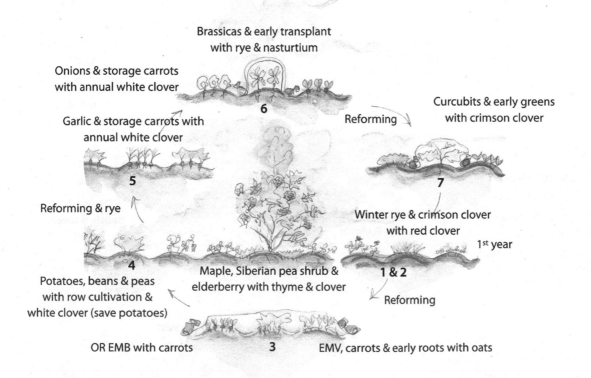

Brassicas & early transplant with rye & nasturtium

Onions & storage carrots with annual white clover

Curcubits & early greens with crimson clover

Garlic & storage carrots with annual white clover

Reforming

6

7

5

Reforming & rye

Winter rye & crimson clover with red clover

1st year

4

1 & 2

Potatoes, beans & peas with row cultivation & white clover (save potatoes)

Maple, Siberian pea shrub & elderberry with thyme & clover

Reforming

OR EMB with carrots

3

EMV, carrots & early roots with oats

f. This rotation leaves the plot with more fertility and organic matter and few weeds. Beds in mixed brassicas are left as crop-cover crops and overwinter trap crops and habitat hotels.

2. Cover Crop/Summer Mixed Veg (2nd year)

a. Management specifics: beds in the previous summer mixed brassicas are power harrowed, stale seeded and seeded into summer mixed vegetables. Row cover is only used if needed. Sprinklers are used for irrigation.

b. GUTs: clover/spinach/clover and salad/clover/cilantro

c. Cover crops help to shade and cool summer greens.

d. This is great for July spinach, lettuce, radish, turnip.

e. Also mulch produced from cover crops can be used for bedding down overwintered spinach and hardy kales.

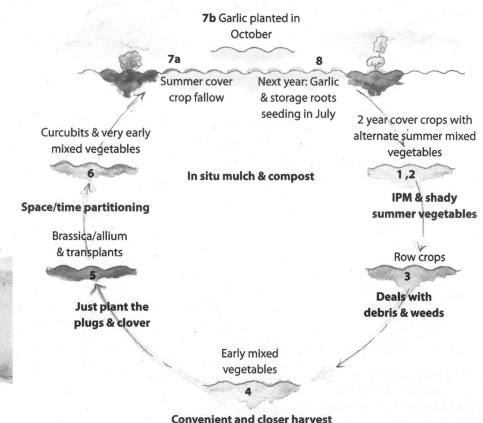

Guild Crop Rotation Services and Themes

This is our current state of our guild crop rotation's evolution.

NOTE: we have provided some companionship themes in **bold**.

7b Garlic planted in October

7a
Summer cover crop fallow

8
Next year: Garlic & storage roots seeding in July

2 year cover crops with alternate summer mixed vegetables

1,2

IPM & shady summer vegetables

Row crops

3

Deals with debris & weeds

Curcubits & very early mixed vegetables

6

In situ mulch & compost

Space/time partitioning

Brassica/allium & transplants

5

Just plant the plugs & clover

Early mixed vegetables

4

Convenient and closer harvest

f. Mixed vegetables can be left as crop-cover crops.

g. This rotation leaves some weeds and trash to be easily dealt with by row crops.

3. Row Crops

a. Management specifics: drip irrigation is moved from brassicas/alliums; hilling is mechanized if appropriate to scale. Trap crop bed is planted in previous year's row (for Colorado potato beetle and leek moth; eggplant, potatoes and garlic are used as trap).

b. Row crops tolerate more debris left over from two years of cover crops and any weeds from mixed veg beds.

c. GUTs: Peas/potatoes/bean and potatoes/peas/potatoes

d. Row crop beds can be reformed and cover cropped as beds become available.

c. They leave a weed-free, trash-free, raised and moderately fertile environment for EMV.

4. Early Mix Vegetables (EMV)

a. Management specifics: Beds are harrowed, false sowed, seeded and covered with row cover. Irrigation: sprinklers are used, paths are cover cropped.

b. GUTs: Carrot/salad/beet and spinach/carrot/cilantro and carrot/salad/carrot

c. This a big space available for getting a lot of those popular market garden crops seeded in a suitable site.

d. They leave the garden with cover cropped paths and layers of cover crops and crop-cover crop from a summer of relay cropping.

e. Do not mix any brassicas in here, as it complicates integrated pest management (IPM) because too many guilds have brassicas already.

5. Brassicas/Allium and Misc. Transplants

a. Management specifics: Beds are harrowed and mulched with rows of drip tape; beds are dibbled, planted and hooped with well-sealed row cover.

b. GUTs: Broccoli/head lettuce/onion and kale/beets/onions and cauliflower/parsley/fennel and onions/chard/cabbage

c. Most of these are heavy feeders and require compost application and micronutrients.

for beets and broccoli especially (boron deficiency is an issue).

d. Path cover crop will survive and fill in the paths with white clover

e. This guild crop triad leaves major debris and uses fertility

f. Beds should have drip removed, mowed to 1" above ground, reformed and cover cropped into winter rye by early October. Alternate beds can be left bare if plots are NOT vulnerable to spring erosion.

6. **Curcubits and Very Early Mixed Veg**

a. Management specifics: Beds are very shallowly harrowed and composted; mulch and drip tape are installed and beds are either seeded into very early mixed vegetables at the end of April or await Cucurbit transplanting at end of May.

b. GUTS: Melons/salad/squash and cilantro/squash/spinach and zucchini/clover/summer squash

c. This is a rotation that requires a lot of fertility. Obviously squash are heavy feeders and grow for many months, but also very early mixed vegetables has a high fertility demand because the soil is colder and less soil life is active to move nutrients around.

d. This rotation leaves the plot with moderate vine debris, continuing decomposing brassicas root structures and intercropped cover crops.

7. **Summer Cover Crop and In situ Mulch Production ⇢ Fall Garlic Planted**

a. Management specifics: Beds are in rye/clover/rye in spring, clover beds are reformed in summer and seeded into buckwheat.

b. Buckwheat beds are mowed before seeding and prepared for garlic.

c. Garlic beds are mulched without drip tape, second-cut hay is a good option, use of in situ rye mulch is possible.

d. Rye beds could be sickle mowed or flail mowed and set the debris on adjacent beds.

e. This guild crop triad leaves the plot well established for winter with alternating beds of mulched garlic and undisturbed rye residue.

f. Alternatively, rye could be rolled/crimped and furrows could be made for planting garlic in the in situ linear mulch. Also, rye beds could be left to self-sow rye or be seeded into oats.

8. **Garlic and Storage Carrots**

a. Management specifics: Garlic start to grow early in spring and is monitored for pests. Predatory wasps and row cover are used for pest management of leek moth. Sprinklers are available if needed.

b. Self-sown rye beds (or oats) are reformed and cover cropped into oats in spring, and flail mowed and power harrowed for fall storage roots. Carrots are row covered to prevent carrot rust fly.

c. Paths are seeded into white clover, garlic beds are seeded into rye and red clover. Carrots beds are seeded into rye.

d. This guild crop triad leaves the plot with a lot of high-carbon material from garlic stalks, mulch, as well as other carrot and garlic debris. The beds are covered with nitrogen-fixing legumes to start next year by balancing the high carbon for compost production.

PermaInvestment

INVESTING IN YOUR FUTURE

Once we begin down a journey of managing an efficient and diverse market garden, we can see the future opening up ahead. Our possibilities are many, because our investments are numerous when we integrate perennials, annuals and/or animal agriculture. We can continue to evolve our production toward a desired agro-ecology: woodland or other. We can transition our guild enterprise production to encompass new crops and product ventures or we can rely on the resiliency we have built through diversification of our income stream and reduced vulnerability to climate and weather.

But what about all those ecosystem services? Are we really getting enough immediate return to justify these perennial investments?

First of all consider:

Pattern propagation is good math

You can buy 25–100 elderberries (age depending) for $100 and within 3 years begin serious propagation from cuttings rooted in water.

Second remember:

Annual return on perennial investment

Market gardeners often avoid perennial expenses because they don't yield for years to come and perennial-based farms (orchards, berry farms) often lack the diversity to truly benefit from perennial agriculture's many services (other than food).

By investing annually in perennials, we will be well placed for a resilient future. But that is not enough. There are so many costs that demand our wallets and time. We must see *annual returns from perennials*! In this way, we have a tight package to present for budgeting that will win us the right to invest more fully in our farm's future.

Most farm investments (seed, supply, small tools) will see return within 1–3 years. Medium-sized investments (irrigation, tractors, equipment) will see return within 3–6 years. Larger infrastructure will take more time still. We can certainly justify the return on perennial (ROP) as being similar to a 3–6 year investment; many are already producing marketable yields in this time frame. But I would argue that there is return within 1–3 years for many perennials, including larger, emergent trees that won't yield crop for decades.

GREEN THUMB TECHNIQUE

Pattern Propagate Successful Species

Use our three stage technique for investment in perennials:
1) plant index guilds in hugelkultur composts,
2) plant research guilds in nursery gardens,
3) plant production guilds in permaplots.

Examples of ways perennials can return annually within 3 years.

- Improved soil and soil life
  - Perennials enhance soil aggregation, feed soil life and shelter the soil environment.
  - If your soil life is teeming and ready to go come spring, your crops will grow faster and taste better.
  - Improved nutrient cycling with active soil food web.
  - Improved crop root growth with loose soil structure.
- Beneficial organism husbandry
  - Conserve and build populations of predatory insects to help keep pest populations in check. We can do this by creating habitat and providing food sources for these throughout the year.

Growing the Rainbow

Beets

Carrots

Radish

Chard

Beans

Potatoes

Onions

Cherry tomatoes

Peppers

Zucchini & summer squash

Resilience is its own form of insurance!

- Windbreak, snow fence, soil retention
  - Your fields will be much healthier, holding snow on the land so it can feed your aquifer, buffering overwintered and early crops against howling winds and prevent erosion and nutrient loss. Your fertility cost will be reduced, and your soil productivity will increase.
- Nitrogen fixing, nutrient mining and sequestering
  - Your soil is increasingly enriched from the air and subsoil.
  - Leaf fall, organic matter additions, subsoil nutrient availability and nitrogen fixation by diverse perennial species will reduce your fertility costs.
- Nursery
  - You can start selling divisions, cuttings and seeds of many species within only a few years.
- Agro-tourism/research
  - There is a great attraction to farms with diverse undertakings, especially those showing how we can profitably cultivate an ecosystem farm.
  - Reduced marketing due to attractiveness of operation.

Consider these further returns on perennial investment

Consider the marketing magic of mixing annual crops in bunches. Consider the increased benefit of adding other complementary products to your table.

- Beauty, job satisfaction and property value
- Research value, education opportunities
- The sky is the limit when it comes to the knowledge locked in diverse agro-ecosystems.

JUST REMEMBER: Stick to your guild enterprise production and holistic goal so you don't become spread too thin.

Plant diverse perennials in index guilds, choosing the best for research nurseries and pattern propagating them in permaplots.

Profit Resilience

Assess

Management — You — Ecosystem

Community — Land

Design — Production — Service

Hugelkultur Index Guild

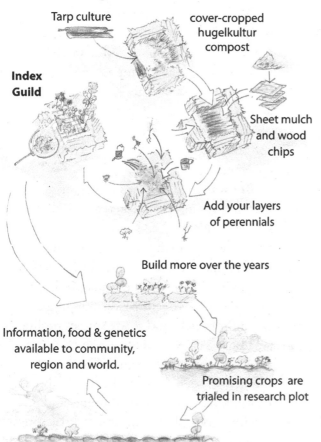

Tarp culture

cover-cropped hugelkultur compost

Index Guild

Sheet mulch and wood chips

Add your layers of perennials

Build more over the years

Information, food & genetics available to community, region and world.

Promising crops are trialed in research plot

Best varieties are planted in site-appropriate PERA triads.

Resilience Kit

These are solid actionable takeaways... ...Get 'er done!

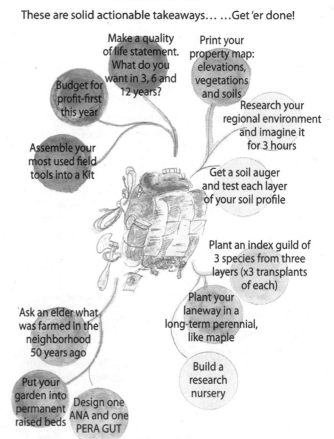

Make a quality of life statement. What do you want in 3, 6 and 12 years?

Budget for profit-first this year

Print your property map: elevations, vegetations and soils

Research your regional environment and imagine it for 3 hours

Assemble your most used field tools into a Kit

Get a soil auger and test each layer of your soil profile

Plant an index guild of 3 species from three layers (x3 transplants of each)

Ask an elder what was farmed in the neighborhood 50 years ago

Plant your laneway in a long-term perennial, like maple

Put your garden into permanent raised beds

Design one ANA and one PERA GUT

Build a research nursery

SUCCESSION SUCCESS

The ultimate outcome of permabed production is a natural evolution of your garden toward agro-ecology, regenerative productivity. You will gradually have increasing yields of perennial and self-seeding annuals and biannuals. But you will also have increasing networks below the soil of mycorrhizal, healthy populations of rhizobia and other teeming soil life. The trees will be annually shedding organic matter, and chosen crops, animals and other productions will flourish within your guild enterprise production.

Ecosystems eventually reach a steady-state equilibrium, where they maintain themselves somewhat in terms of species diversity and net primary productivity. As stewards of the land, we can choose facilitated permaplot enterprise patterning to maintain some plots as open annual ecologies or fruit woodlands or turn some into pastured savannah.

We can leave plots to mature into old-growth forest and receive myriad benefits. They provide exceptional habitat for beneficial organisms and endangered species, as well as selective harvest of timber and firewood. Needless to say, there is beauty.

But whoa! What is that? This is an old-growth forest that was planned and organized! It has such desirable species in wild abundance as oak for acorn-fed pigs, chestnut and walnuts for specialty markets and maple and birch for syrup. It has edible mushrooms that have been introduced and thickets of currants, gooseberries and ephemerals.

But wait. This isn't wild foraging in a bush lot; when I say thicket, I mean berry laneway. This is an edible ecosystem that has been organized: access is simplified through maintenance of alley beds. Desired sun-loving species can be managed through maintenance of open spaces, chipping the wood for mulch, and shady species proliferate under a layered edible canopy.

As the years go by, you can manage for diversity of plots. Expand PERA triads to PERA plots with patterned propagation or redefine ANA plots with timely disturbance.

Agro-ecological Evolution

Annual-dominated production

Begin integrating perennial islands as PERA triads

PERA triads begin to dominate, patterned propagation is used to move genetic material to new triads

A chosen steady-state agro-ecology emerges and is managed using the permabed or other system

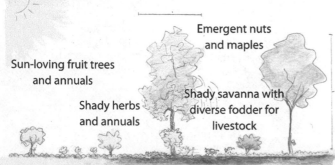

Sun-loving fruit trees and annuals

Emergent nuts and maples

Shady herbs and annuals

Shady savanna with diverse fodder for livestock

Distinct ecologies exist for our productions

PERA triads can become entire plots of perennials with alley beds. ANA plots can be broken up with berry hedges in places or left as highly productive open space for common market vegetable favorites.

> *You can continuously evaluate, choose and design manage for the edible ecosystem you want.*

Maintaining Market Garden Favorites

What's that? How will we be able to maintain current production of market garden crops? First of all, within 3, 6 and 12 years, you will be bringing other desirables to market: elderberries, black raspberries and soon, pears or hazelnuts. And you may be venturing

into small livestock, edible fungi, etc. — whatever you have chosen for your evolving guild enterprise production. However, you need not forsake market favorites, you can maintain your beds for annual crops that are preferred at your markets. You can maintain open spaces for sun-loving annuals through reforming beds and using conservation tillage. Planned tillage for an ANA permaplot will prune tree roots near the surface and prevent them from encroaching. Perennial field patterning prevents a full canopy closure unless pattern propagation is used to turn a permaplot into a block of forest.

Also, as your agro-ecology evolves, you can select for more shade-tolerant annuals, biannuals and perennials for ANA triads adjacent to PERA triads and place the sun lovers where you maintain openings for their ideal production. The list on page 250 shows how many market favorites can tolerate less sun without a loss in quality. I'll take slower maturity in some places in order to maximize my photosynthetic strata.

 Do You Know?

Increased Ecological Exchange

As your garden is managed for agro-ecological succession and soil life is conserved through bed reforming, there will be a marked increase in below-ground diversity, nutrient movement and soil life recolonization following minor bed preparation work.

> *Permaplot succession* means a movement toward a maturing agro-ecosystem where disturbance plays a role to maintain plot space for annuals to flourish.

I want to maximize the layers of desirable species between healthy soil and the sun.

Furthermore, at this stage in your agro-ecosystem's succession all the benefits — soil health, organic matter, windbreaks, summer cooling, etc. — that come from planned integrated production will outweigh any loss of annual production space. In addition, your preferred annual crop guilds will benefit more and more from ecosystem complexity.

Management Points

Important to your end goal is managing for access. Your emerging agro-ecology can quickly become a choked jungle even before it looks like a forest or woodland. You must adhere to garden patterning principles such as:

- Proper ratio of PERA and ANA triads
- Making sure to make distinctions between emergent PERA triads and regenerative

PERA Triad Access Evolution

Annual crops may be planted with establishing perennials and in adjacent beds

Crops like asparagus can serve as an alley because you can drive over them easily

Sheet mulch

Annual crops can continue to be planted in adjacent beds, especially those benefiting from the garden microclimate

Chip mulch

PERA triads can evolve into a 5-bed pattern and incorporate more perennial ground covers in adjacent beds

Beds adjacent to key PERA bed may be left for ground covers

Perennial crops can be moved out with time to give more space to access key PERA crops

PERA Triad Evolution

Young sapling planted in improved bed

Annual crops   Raspberry   Currant

Raspberry canes harvested for patterned propagation

Emergent species planted in shade & protection of dwarf species

Add thyme ground cover

Cover crops improving future PERA bed

Herb layer can be regenerated and moved if needed

Ground cover alley bed

Alley beds can be planted to crop in early season

PERA triads for extended flexibility of harvest options.

- Also, make sure you maintain alley beds adjacent to any perennial crops through species selection and mowing.
- Use alternate maturity patterning to make sure you can easily manage weeds or encroaching perennials in each season.
- Use edge beds to manage for undesirables, like wild parsnip, invasive grass and others.

This diverse agro-ecosystem has a pattern facilitating management

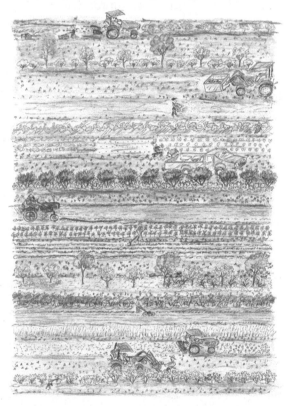

Agro-ecological production is the integrated (commercial) production that considers living and nonliving aspects of an agro-ecosystem and the aim of regeneration of systems through the production process.

Permaplot Evolution

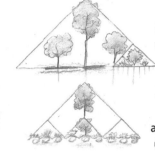

PERA beds are organized in row with the rule of 3

PERA triads will eventually spread their influence to adjacent ANA triads using the rule of 3.

Shading will change the garden ecology with time. This is easy to manage. Adjustments are made triad by triad.

The south remains sunny

North

Sun-loving species

Shade-tolerant species

GREEN THUMB TECHNIQUE

Patterned propagation allows you to choose to proliferate successful species to fill in entire permaplots, transplant into other farm DMZs and to pot up for sale. Use patterned propagation to affordably shift your production emphasis according to your guild enterprise production and based on real field success trials.

POETIC EVOLUTION

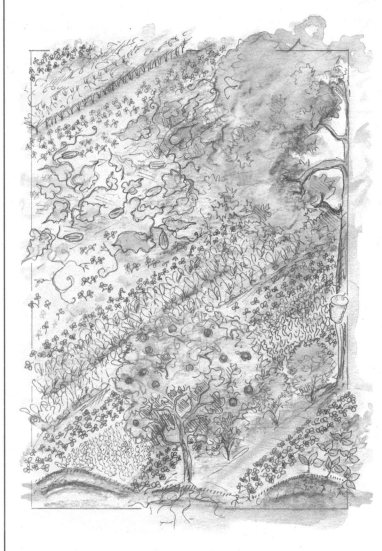

The emergent trees: the oaks, maples, chestnuts will form a canopy overhead.

The fruit trees: the pears, apples and plums will grace the land with well-filtered light reaching to ripen their fruits.

Shrub trees will grow dense and proffer up their hazelnuts, cherries and Saskatoons in tidy rows approachable for easy harvest.

The berry bushes will blush in season with blues, reds and oranges. Fresh currants and gooseberries in the shade, haskap and raspberry in the sun.

The vines will curl up the dwarf apples and sprawl across last year's stems of herbs, hardy kiwi and grape, clematis and beans and peas.

Mints will grow rampant on edges, and curling asparagus fronds will remind you of their tasty spring treat.

Anise hyssop, Saint John's wort and hollyhock will grow amongst.

Beds of cultivated vegetables will be tended between these wild food zones.

Chicken tractors can move through, pigs can wallow, and cattle can graze.

Edge beds are permabeds that are managed along the peripheries of different agro-ecosystems to maintain a desired production type. This includes beds adjacent to pasture lanes, between PERA islands and ANA plots and areas maintained for access along fence rows, trellised crops and fruiting hedges.

Conclusion

FINDING OUR FUTURE

We can sift agricultural history to find grains of wisdom, gleaning from thousands of years of farming to find sustainability for the agriculture of tomorrow. Have we forgotten that agriculture operates within environmental constraints and possibilities? We may as well press the reset button here and go back to basics as a template for improved production: the *ecology of agriculture.*

Furthermore, farming is often stagnating in dogmatic duplicity: neither the *monoculture* farm (a sparse business-as-usual approach) nor the *hippie* farm (an inefficient/disorganized abundance) will suffice. It is the middle ground that will enjoy profit resilience and provide wellbeing for individuals, communities and our greater ecosystems. We must balance the mimicry of natural systems with the practice of profit.

This book is part of a greater pantheon of thoughts, research and aspirations for improved market gardening and sustainable farm practices. If there was anything left unsaid, details missed or misplaced, then let those omissions be the points for continued dialogue. Our model is designed to be adopted and adapted. Through design management we can evolve in sync with an agriculture that is meant to change — an agro-ecological production.

Small farms are the starting point for an agricultural re-evolution. They are situated amidst the Finger Lakes or along Lake Champlain's clay valley; stretched atop glacial moraines or vast lowlands; at the base of the towering Monteregian Hills or nestled in the cragged valleys of the Green Mountains. Small and flexible, these farms can profitably be prototypes for the systems of the future. Some, already proficient in diversified agriculture, may only lack a unifying and flexible system for integrating production and simplifying management. Organizational land patterning facilitates gradual layering of appropriate perennial, animal, and other long-term productions without jeopardizing current productions.

Every market garden is a research plot for ecosystem farm design. The trials and successes of profitable farms using guild enterprise production will serve to define the very boundaries of agro-ecologies. Indeed, soon farm geographic addresses, along with successful crop guild species and other environmental data can be overlaid to classify local agricultural potential. This synthesis of information helps delineate agro-ecologies. These are agricultural ecosystems for the future, whose parameters are defined by the complement of profitable food guilds that are regionally ecologically fit, market-desirable and production-feasible. They must be locally adapted, economically viable, management

Creating an agro-ecological landscape

efficient. A farms successful complement of species is thus the data for a master list, an agro-ecological database, a hub for would-be farmers to input their personal, community and environmental data and review potential agro-ecological enterprises.

Immutable Principles

We need these farms, and they need ecosystem services. These farms, are built around immutable principles, such as building soil, enhancing intercrop services and organizing around time/space/energy management. We can pattern for agro-ecological data to map the future of food. We can do it from the great prairies to the Mississippi Delta and the Rocky Mountain meadows, all the way to the lava folds of Hawaii or the Deccan Plateau of India, and from the rich loess of China to the Murray-Darling Basin in Australia.

Ecosystem farming regenerates our resources, conserving our waters and purifying our air. Whether the Yangtze, Euphrates or Rhone Rivers; whether the Ogallala, Amazon Basin or the East European Aquifer, we must preserve our water resources through improved farming.

No More Aral Sea! No More Desert!

No longer shall we tax our natural resources until the point of disaster. No more Aral Sea! Not like Lake Poopo nor the mighty Colorado River that never meets the sea. A river that doesn't meet the sea is somehow the saddest mystery. Yet, there is no mystery.

Indexing Agro-ecological Landscapes

It's a design process with humble begininings

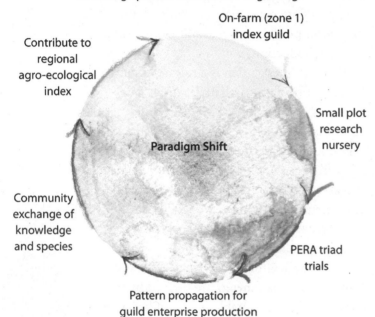

On-farm (zone 1) index guild

Contribute to regional agro-ecological index

Small plot research nursery

Paradigm Shift

Community exchange of knowledge and species

PERA triad trials

Pattern propagation for guild enterprise production

Maintain with minor disturbance

Agriculture

Transition

Agro-ecology

With some old-growth islands

Stagnation of succession, from routine annual tillage

Succession success

We overconsume resources by keeping our farmland in a perpetual state of ecosystem stagnation through constant disturbance. This denies farms the resilience built through ecological succession, which would allow them to regenerate resources for their own productivity. We overuse warm climates in a global agriculture, ignoring the local potential.

The deepest soil in our fields is under the old oak tree left in the middle!

Designing Agro-ecological Landscapes

We all have something we can make into something more

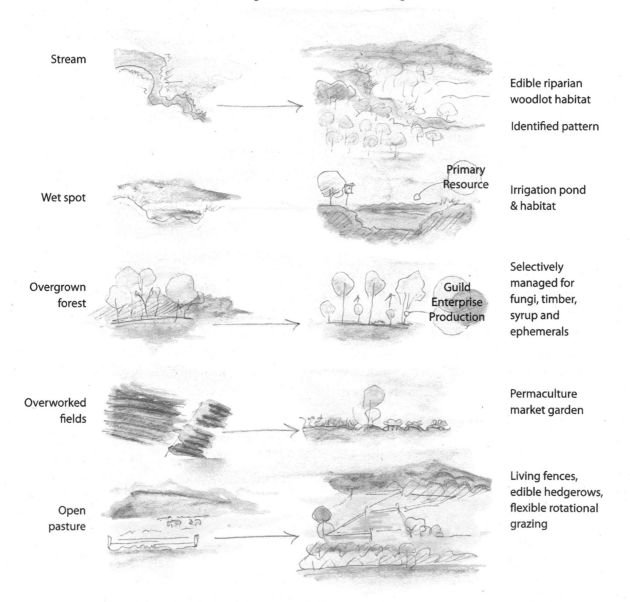

Stream → Edible riparian woodlot habitat

Identified pattern

Primary Resource

Wet spot → Irrigation pond & habitat

Overgrown forest → Guild Enterprise Production — Selectively managed for fungi, timber, syrup and ephemerals

Overworked fields → Permaculture market garden

Open pasture → Living fences, edible hedgerows, flexible rotational grazing

Commit to Ecosystem Services

Commit to the evolution of your farm's landscape along a natural trajectory and so build and stack ecosystem functions and services for your immediate and long-term benefit. Allow soil mycorrhizal networks to weave a gossamer of connectivity under the waves of your garden beds by reforming permabeds routine tillage. Allow your land to have green growing or dead matted material across its surface by prioritizing grow-and-mow weed management. Build profit through overwintered crops like rye, spinach, kale, alongside perennials, all of which maximize winter cover of the land.

Creating soil health

Humility for Humanity

We have only been on this Earth for the shortest of time. Our natural environment is more resilient and mighty than we. We do not change production because of the "environment" but because of "humanity." Because we wish to prosper forward. Rethink how you invest. Invest in your future by

We need an environmental ethic that will tell us as much about using nature as about not using it.

— William Cronon

supporting the agro-ecology that is our only home.

The time is ripe to re-inhabit the land with fierce desire to make a good living off the farm, invested in community, family and soil. Will those who live where they work care for the future of their resource base? Do those who eat from downriver take care to reduce the effluent fed into their food shed? Will the growers produce clean food for their customers? Can we rebuild community food-scapes? Let's put a face on food and feed our many faces? We are an agri-culture!

Pattern your lawn for food. Build micro and macro food businesses. Evaluate your local park and start a community garden. Sit on a council meeting and plan out an edible promenade. Approach policy frameworks with new mindframes. Assess your market garden and plan for the future. Opportunities abound, and the horizon is grand.

If our soils were more biologically active could we sequester more carbon than we

Bed-by-bed in an organized and patterned way we commit!

emit? If our food was tastier, healthier and more available could we alleviate health care? If our economies grew around food, would our economies grow sustainably?

Can you see the great edible tree-studded savannahs of the Midwest, the micro-market gardens of the Catskills, the diversified cash crop meadows of the St. Lawrence lowlands, with hazelnut, elderberry and sunchokes alongside our common staples? Can you see kitchen gardens springing up amongst our urban centers? This is all achievable. We simply need to organize and choose our diversity for best design management. This is possible, because it is already happening!

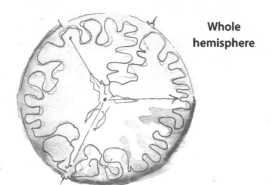

Whole hemisphere

The Leaders Will Be at Every Scale

The leaders will be at every scale, combining wheat, hilling in their Farmall and double digging with their broad-forks. Design for your success and your situation! Account for the life you want and be profitable.

Budget for profit first. Invest a little each year in soil and perennials. Build resilience.

Agro-ecologically fit species make their mark on the map

Unprofitable farms become parking lots, unprofitable farmers are exhausted. So focus on simple diversity; Choose your profit centers and build your business like a guild — simple, slow and sound, and integrated with integrity.

This zeitgeist is more than a zephyr; already the work being done by growers, academics and everyday consumers, is changing the way we perceive food, our community and land. We actually all want to live in a vibrant Garden of Eden, it's evolutionarily hard-wired. Let's make it!

Quite frankly, nature always succeeds, and with a little ingenuity we can guide the succession toward a complement of beneficial species appropriate to the site. Permabeds and other strategies set down in this book can be another spark to kindle the flame of your farm and your journey.

Let's do it, one bed at a time, one farm at a time, one watershed at a time, one nation at a time, together. This is our brave new horizon; let's follow the road maps toward profit resilient agro-ecological farming. Put your dot on the map!

Education is the kindling of a flame, not the filling of a vessel.

— Socrates

**Here is our neighborhood — let's pattern it for agro-ecology
Place your dot!**

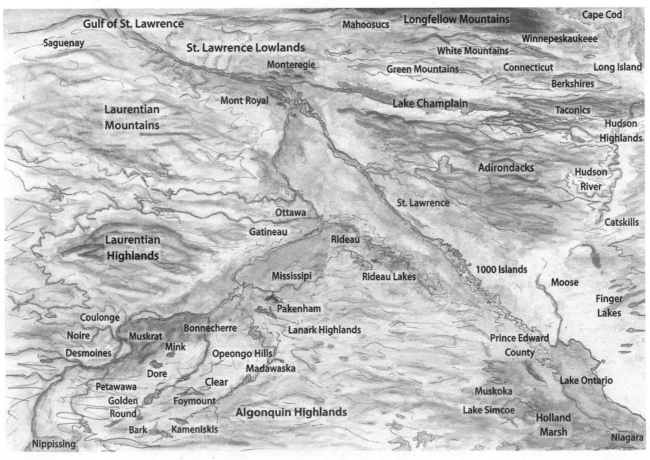

Glossary

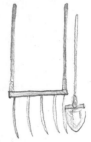

A

Alley beds: Raised beds that allow cart/tractor access to adjacent beds by time-sharing with early harvest crops or use for only low-growing crops.

Alternate maturity patterning: The alternating of garden beds between crops of early and late DTM to enhance grow-and-mow management, access and cover crop priority.

Agro-ecological: Relating to an agriculture that is embedded in the local environment and functions with ecosystem services.

Agro-ecological indexing: The process of diversified farms serving to outline local and regional agro-ecological production potential. The boundaries of these agro-ecosystems are delineated through the overlaying of successful farm addresses. Farm success is correlated to the production's ecological fitness, management efficiency and market-readiness.

B

Bed history: The records of a specific bed's crop rotation, inputs, etc.

Bed-specific character: A bed's specific environment: soil, hydrology, etc.

Beneficial organism husbandry: The enhancement of wild populations of beneficial insects through their introduction and/or habitat provisioning.

Best management (production) practices: Farm practices that meet farmers' needs with improved environmental impact.

C

Casting, worm (vermicast): Biologically rich and nutrient dense worm excrement.

Cation exchange capacity: The ability of a soil to hold and exchange cations (positively charged atom or molecule) with the soil solution, increasing their availability for soil organisms and plants.

Compost path (compostipath): Use of permabed paths to make compost, balancing crop debris, nitrogen fixation and soil life activity.

Conditional germination zone: the area and conditions necessary for seed germination, including mulching, row cover and/or moisture management, to achieve germination without drilling seed into the soil.

Critical crop analysis: The direct comparison of different crop characteristics (habit, form, function), using a crop matrix to correlate with other crops for companionability.

Crop-cover crop: Using a crop as a cover crop once it has been harvested to provide further services: habitat, shading, trap crop, organic matter, etc.

Crop guild: A selection of crops associated for mutual benefit.

Crop guild triad: A guild organized in a triad and assigned to a specific bed (outer, center, outer).

Crop services: Those services from crops above and beyond their edible part, includes nutrient accumulation and fixation, carbon sequestration, beneficial insect habitat and pest deterrence, etc.

Crossbeds: Alley beds that go perpendicular to Permaplots.

D

Design area: A spatial designation for proposed future farm projects.

Design management: An emphasis on design for better management through a consistent built-for-improvement design process and principles.

Design management zone (DMZ): A farm area with specific operations and infrastructure working together and connecting to other DMZs.

Disturbance agriculture: Describes productions using routine tillage.

E

Ecosystem service providers (ESPs): Species and populations providing ecosystem services.

Extensive agriculture: Increases the land base, space between crops and time between cropping in order to reduce the inputs and labor needed.

F

False sow: The practice of preparing a fine seed bed for germinating weed seeds for shallow cultivation prior to seeding crop in the now clean bed.

Flow: Includes natural flows of wind, warmth, animals, etc., as well as the designed flow between and within areas.

Flushing: The managed removal of seeds from a bed in self-sowing cover crop through planned periods of germination in preparation for crops.

G

Garden environment mapping (GEM): Garden plot mapping for better production, including hydrology, pedology, ecology, microclimates, etc.

Grow and mow: A method of weed management that emphasizes growing cover crops to outcompete weeds in place of a till-and-kill model.

Guild crop rotation: An annual crop rotation where crops are grouped into crop guilds for enhanced productivity and other benefits to the greater agro-ecology.

H

Hidden yields (see crop services): Non-usual production yields, such as nitrogen fixation, pest-deterrence, soil structure improvement, etc.

I

Ikigai: Japanese concept for balanced, purposeful, successful and enjoyable living. Means "a reason for being."

Immobilization: The immobilization of nutrients occurs when soil organisms and plants consume inorganic compounds to build organic compounds needed for life function. These nutrients are retained in organisms until released through mineralization.

Imprinting: Seeding of specific cover crops following a specific bed situation, say a diseased crop seeded to a red clover as a way of tracking that situation. This allows a living map of unique garden situations for record keeping and crop management in the seasons to follow.

Index guild: Perennials planted in farm zones 1 and 2 so as to test their hardiness and mutual support in close proximity. It also provides an observation window of field-scale plantings of chosen crops in zone 3, 4 and 5, yet seasonal cycles are easily monitored in zone 1.

Integrated production: Agriculture that integrates multiple species in a given area within the same time frame.

Intensive agriculture: Production that increases capital, inputs and labor to gain higher yields per acre, usually with a lower fallow ratio.

L

Limiting nutrient: A nutrient whose absence is found responsible for limiting productivity.

Nitrogen and phosphorous often limit growth in plants.

M

Mineralization: The oxidation or decomposition of organic matter so as to become available for plant uptake.

Model design: A design from another farm that can be adopted and adapted on your farm through the Design Management process.

Mutualist: An organism that forms a symbiotic relationship with another organism in which both organisms benefit (mutualism).

Mycorrhizal fungi: Mycorrhizae form symbiotic relationships with plants, exchanging nutrients and water for root exudates (sugar).

N

Niche: The ecological role, space and timing occupied by an organism within a community.

Nutrient accumulators: Plants that are good at bringing up nutrients from lower soil horizons, often taprooted and apt to accumulate high concentrations of nutrients in their tissue.

Nutrient scavenger: Organism (plants) that pick up excess fertilizer nutrients quickly and release them later, often preventing leaching.

O

Organizational land patterning: The selection of a defined spatial unit for land organization, environment mapping and eventually patterning for a diverse and efficient agriculture.

Organized garden patterning: The repetition of permabeds and their organization into triads and permaplots to pattern for integrated production.

P

Patterned propagation: A pattern-based strategy for dispersing successful perennials through an agro-ecology from an initial trial bed of a new species.

Pedosphere: The natural sphere on Earth that pertains to the soil (many terrestrial areas to a depth of approximately 1 meter).

PERA triad: A three-bed perennial crop grouping in a permabed system.

Permabeds: Permanent agro-ecological beds; raised beds that are made, reformed but never destroyed in a permabed system.

Permaculture property zones: The division of a property into zones of frequency and accessibility for organized production and projects based on their routine visitation requirements.

Pest cocktail: Plant mixture to distract pests and attract beneficial insects.

Photosynthetic strata: The differing vertical layers of plants that help maximize solar energy toward a land's productivity.

Plant character: Includes an understanding of a crops full life cycle, its habit and resourcefulness.

Production factors: Time, space, energy are universal factors involved in production and should be considered for a balanced management.

Profit resilience: The overarching goal for a new generation of farming, acknowledging that a farm must be both profitable and sustainable in order to succeed in the face of socioeconomic and environmental change. It must actively support natural systems and their ecosystem services and achieve profitability through investing in long-term solutions while maintaining short-term gains.

R

Regenerative productivity: The means of production is regenerated by the mode of production because it supports ecosystem services. Crop production practices then enhance soil life and will improve future crop production.

Relationship: A connection or association between two or more species in your garden ecosystem, including companionship or symbiosis.

Relay cropping: When a second crop is seeded into the first crop before it has finished its production cycle. This is especially beneficial for quick turnaround of land into cover crops.

Research nursery: A small plot for production of perennial propagation material and garden trialing of guild relationships before larger field-scale production.

Rule of three: A conceptual tool that emphasizes three as both a minimum and maximum limit for design elements and/or a multiplier for further organization (e.g., 3, 6, 9, 12). This focuses, limits and improves our designs and productions. See **triad**, **crop guild** and **guild enterprise**.

S

Self-sowing: A crop growing to maturity and releasing seeds, germinating and growing without major intervention.

Sheet mulch: A weed barrier and water retention strategy using layers of cardboard and straw or wood chips through which crops are transplanted. Modern sheet mulches include roll-out type for mechanized production.

Site-appropriate crops: Crops that are selected because they work well with the soil, hydrology, climate, etc., of a given space.

Soil assets: Includes organic matter, pore space, soil life.

Soil integrity: The overall state of soil health: soil organism species diversity and population numbers, CEC, soil structure and aggregation, etc.

Stacking functions: This means getting many outputs (services) from any one element (tree).

Succession: Market garden term for routine seeding of a crop to maintain a constant supply, closely linked to crop DTM and seasonal productivity.

Symbiosis: The close evolved relationship between two species where either both benefit (mutualism), one benefits and the other is unharmed (commensalism), or one benefits to the detriment of the other (parasitism).

T

Tarp culture: The use of black tarps to kill off weeds and as a passive form of land preparation.

Triad: Three adjacent beds that serve as an organizational unit, designated as either annual (ANA) or perennial (PERA) in a permabed system.

U

Umbrella management: The organizing of species with similar management needs into triads for time/space/energy efficiency.

Unconformity: From missing rock layers due to episodes of crustal deformation, erosion or sea level extremes leaving gaps in geologic records.

W

Walkabout: A routine observational movement through fields and facilities.

X

Xeriscape: Landscape design for low-water use, popularized in arid regions.

Resources

Agronomic aspects of strip intercropping lettuce with alyssum for biological control of aphids.
Eric B.Brennan. Biological Control, 2013.

Changes in the land: Indians, colonists, and the ecology of New England.
William Cronon. Hill and Wang, 1983.

Dwarf sour cherries: A guide for commercial production.
Robert Harold Bors, Linda Matthews. University Extension Press, 2004.

Blue ocean strategy: How to create uncontested market space and make the competition irrelevant.
W. Chan Kim, Renée Mauborgne. Harvard Business School Press, 2005.

The new organic grower: A master's manual of tools and techniques for the home and market gardener.
Eliot Coleman, Molly Cook Field, Sheri Amsel, Paul Hawken. Chelsea Green Pub. Co., 1995.

Soil biology primer.
Elaine Ingham, Andrew R. Moldenke, C. A. Edwards. Soil and Water Conservation Society, in cooperation with the USDA Natural Resources Conservation Service, 2000.

Marschner's mineral nutrition of higher plants.
Horst Marschner, Petra Marschner. Elsevier/Academic Press, 2012.

Roles of Arbuscular Mycorrhizas in Plant Phosphorus Nutrition.
Sally E. Smith, Iver Jakobsen, Mette Grønlund, F. Andrew Smith. American Society of Plant Biologists, 2011.

Ecological and evolutionary significance of mycorrhizal symbioses in vascular plants (A Review).
D. W. Malloch, K. A. Pirozynski, P. H. Raven, 1980.

Ottawa-Gatineau Geoheritage Project Field Trip Guide.
Compiled by Quentin Gall. November 2010.

Edible forest garden.
Dave Jacke, Eric Toensmeier. Chelsea Green Publishing Company, 2005.

Permaculture: A designer's manual.
B. C. Mollison. Tagari Publications, 1988.

Permaculture: principles and pathways beyond sustainability.
David Holmgren. Holmgren Design Services, 2002.

Start with why: How great leaders inspire everyone to take action.
Simon Sinek. Portfolio, 2009.

Biology.
Neil A. Campbell, Jane B. Reece, Pearson, Benjamin Cummings, 2005.

Introduction to systems ecology.
Sven Erik Jørgensen. CRC Press/Taylor & Francis, 2012.

Ecosystem ecology.
Sven Erik Jørgensen. Elsevier, 2009.

Growing hybrid hazelnuts: The new resilient crop for a changing climate.
Philip Rutter, Susan Wiegrefe, Brandon Rutter-Daywater.

The organic farmer's business handbook: A complete guide to managing finances, crops, and staff — and making a profit.
Richard Wiswall, Chelsea Green Pub., 2009.

Understanding roots: Discover how to make your garden flourish.
Robert Kourik.

The lean farm: How to minimize waste, increase efficiency, and maximize value and profits with less work.
Ben Hartman, Emma Gerigscott.

Holistic resource management.
Allan Savory. Island Press, 1988.

Managing cover crops profitably.
Sustainable Agriculture Network, 1998.

Organic no-till farming: advancing no-till agriculture — crops, soils, equipment.
Jeffrey Moyer. Acres U.S.A., 2011.

Holistic orchard.
Michael Phillips. Chelsea Green, 2013.

The natural way of farming: The theory and practice of green philosophy.
Masanobu Fukuoka, Frederic P. Metreaud, Bookventure, 1997.

Building soils for better crops.
Fred Magdoff, Harold Van Es. Sustainable Agriculture Network, 2000.

Farmers of forty centuries: Permanent agriculture in China, Korea, and Japan.
F. H. King. Rodale Press, 1973.

Plant Nutrient Management in Hawaii's Soils: Approaches for Tropical and Subtropical Agriculture.
College of Tropical Agriculture and Human Resources, University of Hawaii at Manoa, J. A. Silva and R. Uchida eds., 2000.

Landscape Architecture Theory: An Ecological Approach.
Michael D. Murphy. Island Press, 2016.

Small Is Beautiful: A Study of Economics As If People Mattered.
E. F. Schumacher. Blond & Briggs, 1973.

Soil Erosion: A National Menace.
H. H. Bennett and W. R. Chapline. U.S. Department of Agriculture Circular No. 33., Washington, DC, U.S. Government Printing Office, 1928.

http://www.williamcronon.net/writing/Trouble_with_Wilderness_Main.html

Index

Page numbers in *italics* indicate figures.

About the Author

Zach Loeks **is an educator, designer and grower** who specializes in Edible Ecosystem Design through landscaping and education. He consults widely with homes, farms, colleges, schools and municipalities across Canada and the United States, and through many biomes from Guatemala and South Africa to the Yukon and Mongolia.

Zach manages an award-winning farm with diversified food forest products, heirloom garlic, and a hardy tree nursery. His innovations have won three provincial awards and are featured in his first book: *The Permaculture Market Garden.*

Zach is the director of the Ecosystem Solution Institute, which is dedicated to the education, propagation and inspiration of ecosystem solutions for land use transition. The Institute oversees pathbreaking education sites, including an edible botanical garden near Ottawa, Ontario and a suburban food forest in Winnipeg, Manitoba. Zach is passionate about how small actions - strategically linked - can make big change. His inspiring and empowering vision is presented in his latest book: *The Edible Ecosystem Solution.*

Find out more

www.zachloeks.com

@zachloeks
on Instagram, Facebook, Youtube, Twitter

ZACH LOEKS

Edible Eco-system Design

About New Society Publishers

NEW SOCIETY PUBLISHERS is an activist, solutions-oriented publisher focused on publishing books for a world of change. Our books offer tips, tools, and insights from leading experts in sustainable building, homesteading, climate change, environment, conscientious commerce, renewable energy, and more — positive solutions for troubled times.

Sustainable Practices for Strong, Resilient Communities

We print all of our books and catalogues on **100% post-consumer recycled paper,** processed chlorine-free, and printed with vegetable-based, low-VOC inks. These practices are measured through an Environmental Benefits statement (see below). We are committed to printing all of our books and catalogues in North America, not overseas. We also work to reduce our carbon footprint, and purchase carbon offsets based on an annual audit to ensure carbon neutrality.

Employee Trust and a Certified B Corp

In addition to an innovative employee shareholder agreement, we have also achieved B Corporation certification. We care deeply about *what* we publish — our overall list continues to be widely admired and respected for its timeliness and quality — but also about *how* we do business.

For further information, or to browse our full list of books and purchase securely, visit our website at: **www.newsociety.com**

New Society Publishers

ENVIRONMENTAL BENEFITS STATEMENT

For every 5,000 books printed, New Society saves the following resources:[1]

46	Trees
4,205	Pounds of Solid Waste
4,627	Gallons of Water
6,035	Kilowatt Hours of Electricity
7,644	Pounds of Greenhouse Gases
33	Pounds of HAPs, VOCs, and AOX Combined
12	Cubic Yards of Landfill Space

[1]Environmental benefits are calculated based on research done by the Environmental Defense Fund and other members of the Paper Task Force who study the environmental impacts of the paper industry.

MIX
Paper from
responsible sources
FSC® C016245

new society
PUBLISHERS
www.newsociety.com